Library Technology
REPORTS

Expert Guides to Library Systems and Services

Libraries and Mobile Services

Cody W. Hanson

ALA Tech Source
alatechsource.org

American Library Association

Library Technology
REPORTS

ALA TechSource purchases fund advocacy, awareness, and accreditation programs for library professionals worldwide.

Volume 47, Number 2
Libraries and Mobile Services
ISBN: 978-0-8389-5830-8

American Library Association
50 East Huron St.
Chicago, IL 60611-2795 USA
alatechsource.org
800-545-2433, ext. 4299
312-944-6780
312-280-5275 (fax)

Advertising Representative
Brian Searles, Ad Sales Manager
ALA Publishing Dept.
bsearles@ala.org
312-280-5282
1-800-545-2433, ext. 5282

Editor
Dan Freeman
dfreeman@ala.org
312-280-5413

Copy Editor
Judith Lauber

Editorial Assistant
Megan O'Neill
moneill@ala.org
800-545-2433, ext. 3244
312-280-5275 (fax)

Production and Design
Tim Clifford, Production Editor
Karen Sheets de Gracia, Manager of Design and Composition

Library Technology Reports (ISSN 0024-2586) is published eight times a year (January, March, April, June, July, September, October, and December) by American Library Association, 50 E. Huron St., Chicago, IL 60611. It is managed by ALA TechSource, a unit of the publishing department of ALA. Periodical postage paid at Chicago, Illinois, and at additional mailing offices. POSTMASTER: Send address changes to Library Technology Reports, 50 E. Huron St., Chicago, IL 60611.

Trademarked names appear in the text of this journal. Rather than identify or insert a trademark symbol at the appearance of each name, the authors and the American Library Association state that the names are used for editorial purposes exclusively, to the ultimate benefit of the owners of the trademarks. There is absolutely no intention of infringement on the rights of the trademark owners.

ALA TechSource
alatechsource.org

About the Author

Cody Hanson is the web architect and user experience analyst at the University of Minnesota Libraries, where he works to make the online research process more intuitive and fruitful for students, staff, and faculty. He was a 2010 ALA Emerging Leader, and he co-chairs the LITA Education Committee. As an adjunct faculty member at Saint Catherine University, he has taught "Library 2.0" in the Master of Library and Information Science program.

Abstract

For the past thirty-plus years, libraries and librarians have perceived themselves as subject to near-constant technological upheaval and information revolution, largely due to the rise of microcomputing, desktop computing, and Internet connectivity. Tech and industry prognosticators believe that the impact of mobile computing on our society and economy will dwarf these earlier innovations. Key indicators point to profound implications for delivery of information, access to services, shifts in the demographics of connected users, and broadband access business models. Libraries are uniquely positioned to advocate for the responsible evolution of mobile connectivity, and must move aggressively into provision of library services in the mobile realm.

This report examines the meteoric uptake of smartphones and mobile broadband, describing the trends transforming the way users are accessing information, and the implications for library policy and advocacy. Included is a detailed overview to bring you up to speed on the leading mobile operating systems, including Android, iOS, BlackBerry OS, WebOS, and Windows Phone 7, as well as a discussion of the hardware advancements that have helped the humble phone replace your Walkman, Game Boy, camera, and GPS. To help you decide how to proceed with mobile services in your library, the report includes basic strategies for making your library resources mobile friendly and for developing mobile websites and mobile applications.

Subscriptions
alatechsource.org/subscribe

Contents

Chapter 1

Why Worry about Mobile?

Abstract

Mobile devices are ubiquitous in today's society, and there's no evidence that that is going to change. According to the Pew Internet and American Life Project, as of mid-2010, 82 percent of American adults own a mobile phone or a mobile computing device that works as a phone. This chapter of Libraries and Mobile Services *sets the stage for the Report, explaining why it is crucial for librarians to understand mobile devices and provide services through them.*

I've always had a fascination with gadgets, especially small, portable gadgets. I recall debating with my parents about why a Casio Databank watch would be the perfect souvenir by which to remember a family vacation. And as befits someone who was firmly in the target demographic of the 1980s Transformers craze, multifunction gadgets hold a particular attraction.

The aforementioned predilections go only so far toward explaining how I found myself on a Saturday morning in the fall of 2003 waiting in my car for my local GameStop to open. That October week had seen the release of what I was convinced was a ground-breaking convergence gadget: the videogame- and MP3-playing, Web-browsing smartphone. I was waiting for the privilege of exchanging hard-won U.S. currency for a Nokia N-Gage (figure 1).

If you're familiar with the N-Gage, you're likely wiping away tears from derisive laughter. If not, allow me to explain why the N-Gage holds a special place in the hall of fame of misguided, poorly designed gadgets. Sure enough, it played videogames, which were sold on small memory cards. When the time came to swap cards so you could play another of the few games ever released for the device, you needed to power down the N-Gage, remove the back cover, and pop out

Figure 1
The Nokia N-Gage (Photo credit: J-P Kärnä, "Nokia N-Gage," http: commons.wikimedia.org/wiki/File: Nokia N-Gage.jpg, licensed under the Creative Commons Attribution-Share Alike 3.0 Unported license, http://creativecommons.org/licenses/by-sa/3.0/dee.en).

the battery, underneath which was nestled the *Tony Hawk* or *Red Faction* card.

It played MP3s all right, which you could load on a memory card. The only problem was that memory card occupied the same slot as the game card, so switching from the game I'd been playing on the bus to the music I wanted to listen to on the walk home required balancing the cards, cover, battery, and phone on my lap before my stop.

The challenges of using the N-Gage as a gaming and multimedia device had nothing on the indignity of actually talking on the thing. In order to cram gaming controls and a number pad on the face of the N-Gage, the engineers at Nokia placed the speaker and microphone along the top edge. So, rather than holding the flat face of the device to my ear, as one might expect, I talked into the long, narrow edge, with the device protruding from my head like a fin. Imagine trying to talk into the bottom of a hard-shell taco. Now imagine my wife shaking her head as she bemoaned my "giant crazy phone."

My wife wasn't the only one who mocked the N-Gage. "Sidetalking" became quite the meme in 2003–2004, and jokesters shared photos of themselves online talking into the side of improbably large and unwieldy props (figure 2). Visit the Sidetalkin' website for some galleries that immortalize the mockery of my giant crazy phone.

Sidetalkin'
http://sidetalkin.com

Figure 2
Your author, sidetalkin'.

The N-Gage was undoubtedly a commercial failure, and while I didn't love it personally, it did open my eyes to what was possible in a mobile device. It was the first time I had a web browser in my pocket at all times—a slow, clunky WAP (Wireless Access Protocol) browser, but a browser nonetheless. It ran Nokia's Symbian Series 60 smartphone operating system, and there were dozens of free and paid applications available. I had a primitive geolocation app that allowed me to define places like "home" and "work" by the cell towers in the vicinity and change settings on the phone depending on where I was (at work, turn the ringer off; at home, turn the ringer and Bluetooth on, etc.). It was unlocked, meaning it could be used on any GSM mobile carrier. And I have a vivid memory of waiting for the bus one afternoon, listening to music on the hands-free headset I hacked together myself. The music faded down, and through my headphones I heard my ringtone. I pressed a button on the headset and answered a call from my sister. I pressed it again to hang up, and the music faded back up. I felt pretty sure I was living in the future.

I sold the N-Gage after a few months and reinvested the proceeds in my first iPod, a device I'd resisted to that point as being too limited, a unitasker. In the years since, I've owned the first-generation Motorola RAZR, which was the first phone that I could sync with my computer. I bought the BlackBerry Pearl, Research in Motion's (RIM's) first device explicitly targeted at the consumer market, on day one. I'm typing this on an iPhone. I've got boxes full of old phones, chargers, and cables cluttering up my basement, all artifacts of an ongoing search for the perfect mobile device.

Along the way I became a librarian. I realized that my passion for technology and information had a natural intersection in librarianship, and I've spent the past five years working on better ways to marry the two. It's become clear to me over the past couple of years that mobile is the frontier for information creation and access. So you may not have heard my name before. I may not be a universally recognized expert. But I care deeply about this stuff. I've invested countless hours and squandered hundreds of dollars to amass the knowledge that I hope to pass on in this issue of *Library Technology Reports*.

Why Mobile?

Hopefully it's clear that I care about mobile computing. (Perhaps a little too much?) And if you're reading this issue of *Library Technology Reports*, I'd hope that you care about it as well, or at least have a passing interest. One thing that I learned in library school is that for the past thirty-plus years, libraries and librarians have perceived themselves as subject to near-constant technological upheaval and information revolution, largely due to the rise of microcomputing, desktop computing, and Internet connectivity. I'll discuss in this chapter why tech and industry prognosticators believe that the impact of mobile computing on our society and economy will dwarf these earlier innovations. Can we measure or predict exactly how this impact will be felt by libraries? Perhaps not, but key indicators point to profound implications for delivery of information, access to services, shifts in the demographics of connected users, and broadband access business models. Libraries are uniquely positioned to advocate for the responsible evolution of mobile connectivity, and I would argue are duty-bound to move aggressively into provision of library services in the mobile realm.

I'm sure your library is cash-strapped, underresourced, understaffed. Development of tools and services that target mobile likely seems a distraction, a drain on your time and attention. It might feel like it's the flavor of the month, blustery conference paper fodder that's unlikely to pay off in real service to users.

What evidence would provide a good indication that the day had come for your library to focus

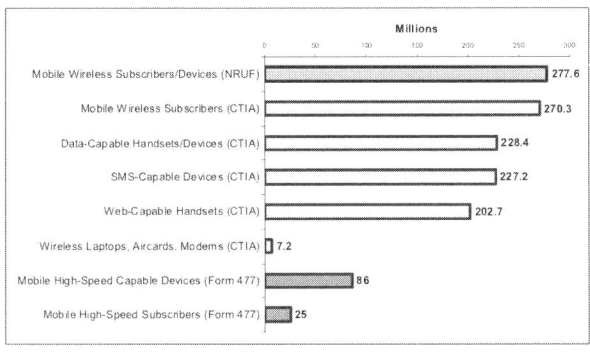

Figure 3
FCC statistics on mobile device ownership in the U.S.

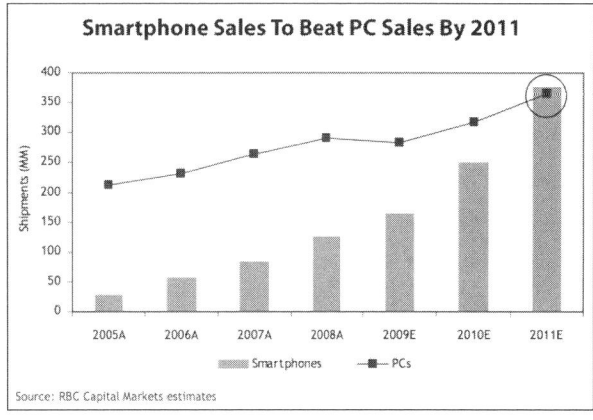

Figure 4
Smartphone sales are poised to overtake PC sales (Original image: Dan Frommer, "Chart of the Day: Smartphone Sales to Beat PC Sales by 2011," *Business Insider* website, Aug. 21, 2009, www.businessinsider.com/chart-of-the-day-smartphone-sales-to-beat-pc-sales-by-2011-2009-8 [accessed Jan. 4, 2011]).

concerted efforts on mobile services? If nearly all Americans owned cell phones? Maybe if a large percentage of those phone owners demonstrably used their device to access the Internet? Perhaps if smartphone sales began to approach sales of PCs? If major information service providers were shifting their focus from the desktop to mobile devices? If the trend turned away from mobile devices mimicking the functions of desktop computers, and instead desktops began to emulate mobiles? Maybe if there was evidence that traditional desktop connectivity wasn't reaching people who could be reached on their mobile devices?

If so, then that day is today.

Mobile Device Ownership

The Federal Communications Commission (FCC) counted over 270 million mobile phone users in the United States in 2009.[1] According to the Pew Internet and American Life Project, as of mid-2010, 82 percent of American adults own a mobile phone or a mobile computing device that works as a phone (this statistic does not include laptop computers).[2] The percentage is higher for younger adults. Fully 90 percent of Americans aged 18–29 own a cell phone.[3]

The Educause Center for Applied Research (ECAR) surveys undergraduate students at U.S. colleges and universities each year about their ownership and use of technology in its *ECAR Study of Undergraduate Students and Information Technology.* In 2004, 82 percent of study respondents reported owning a cell phone.[4] In subsequent years that number rose so high that they have since ceased to bother asking.[5]

Smartphone Sales

By the time you read this, the day that smartphone sales surpass PC sales may already have passed (figure 4). Industry watchers have variously pegged this inversion to occur anytime between 2010 and 2012. An RBC Capital analyst quoted by a *Fortune* magazine blogger predicts that between 2009 and 2012, the number of smartphone users worldwide will more than triple, from 165 million to over 500 million.[6]

The FCC reported that at the end of 2009, 42 percent of consumers owned a smartphone.[7] According to the FCC, smartphones accounted for 44 percent of total mobile phone sales in 2009, and for 50 percent of phone upgrades.[8] The FCC's definition of a smartphone is restrictive, however, as it counts over 228 million active mobile devices capable of receiving data service, and over 202 million active devices capable of browsing the Web.[9] Over 35 percent of U.S. adults now own phones upon which they can install software applications, according to Pew. That number does not include devices like the iPad or iPod Touch, which can use apps, but which don't function as a traditional cell phone.[10]

ECAR found that fully 62.7 percent of U.S. undergraduates report owning an "Internet-capable handheld device."[11]

Mobile Internet Access

Users are taking advantage of the nonvoice connectivity provided by the current generation of mobile devices (figure 6). The Pew Internet and American Life project tells us that in 2010, 40 percent of adults in the United States report using their phone for Internet access, to send or receive e-mail, or for instant messaging. In 2009, that number was 32 percent.[12] A significant majority of young adults access the Internet from

Library Technology Reports   alatechsource.org   February/March 2011

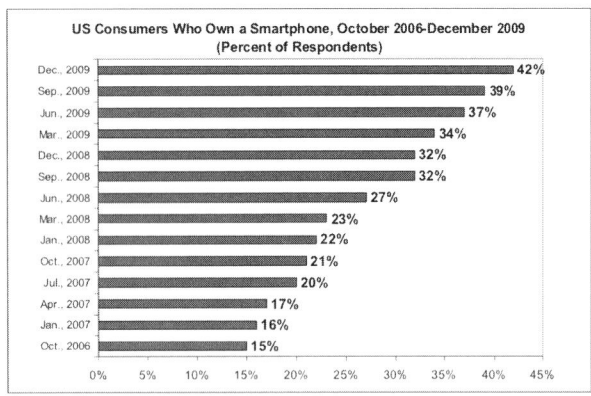

Figure 5
FCC statistics on smartphone ownership.

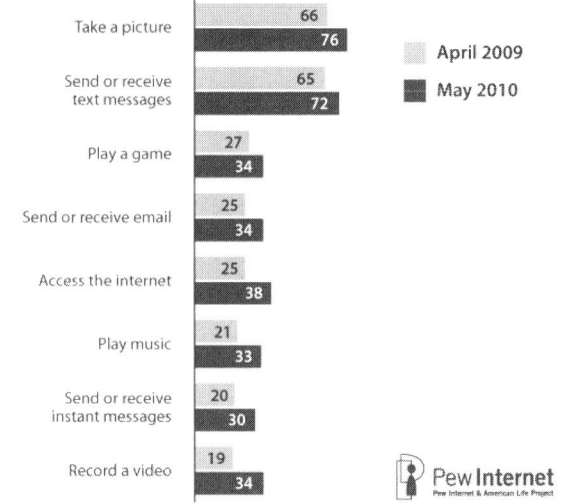

Figure 6
The use of nonvoice phone applications has exploded in the last year.

their phones, a full 65 percent according to Pew.[13] A comScore press release reported that just over 30 percent of American cell phone owners age 13 and older used their device's web browser in an average month in early 2010.[14]

Those mobile Internet users are not just casual users. As of May 2010, Pew reports that 55 percent of mobile Internet users were going online with their phones at least daily, and 43 percent used the mobile Internet several times a day (figure 7).[15] ECAR found that the frequency of mobile Internet use among undergraduates spiked in 2010. In 2009, just under half of undergrads who owned Internet-capable handheld devices used them to access the Internet weekly, and 29 percent did so daily. In 2010, 66.6 percent were online weekly, and 42.6 percent were daily users of the mobile Internet.[16]

I suppose it should come as no surprise then that between 2007 and 2010, AT&T Wireless, the U.S. mobile service provider for the iPhone, reported that demand for mobile bandwidth increased by nearly 50 times. You read that right. A 4,932% increase.[17]

Mobile versus the Desktop

I can hardly imagine the gigabytes of server space devoted to archiving library-related e-mail discussion threads in which debates have raged over the precise extent to which Google is good or evil, benevolent or creepy, friend or foe to libraries and librarianship. Regardless of your particular stance on these issues, I think it's clear that if we want to stay abreast of innovations in information services and delivery, we would be foolish not to watch Google closely.

Google has of course developed its own mobile operating system, Android, which I'll cover in more detail later. Beyond Android-specific mobile applications and services, Google has been very aggressive in creating best-of-breed web tools for a wide variety of mobile devices. And according to its CEO, Eric

Schmidt, mobile is now considered the search giant's first priority when developing new services. Speaking at Mobile World Congress in early 2010, Schmidt said that his engineers are now working on mobile implementations first, before versions optimized for desktop browsers. "We understand the new rule is 'mobile first' in everything. . . . Mobile apps are better apps," he said, according to eWeek.[18] PCMag quoted him as saying, "Every product announcement we've done recently—of course we'll have a desktop version—but we'll also have one on a high-performance mobile phone."[19]

The success of Apple's iOS device platform, the operating system that powers the iPhone, iPod Touch, and iPad, has exceeded the expectations of almost everyone outside the company's Cupertino, California, headquarters. iOS was born out of, and shares a great deal of underlying code with, Apple's OS X desktop operating system. At a recent event, CEO Steve Jobs outlined the ways in which Apple's mobile development is now informing their desktop computer operating system. The event was called "Back to the Mac," and at it Jobs and other Apple executives described how they were bringing the wildly successful App Store model from their mobile devices back to OS X 10.7, the new version of the Macintosh OS. Directly informed by the immersive full-screen applications developed for the iPad, many core applications in OS X 10.7 will take over users' screens, eliminating the menu bars and windows we've long associated with multitasking desktop computers.[20]

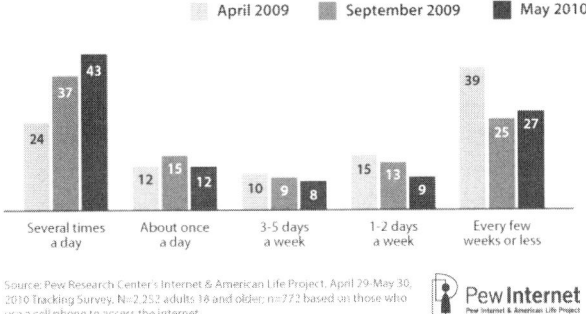

Frequency of cell phone internet use among those who go online from a cell phone (% of adult cell phone internet users)

Source: Pew Research Center's Internet & American Life Project, April 29-May 30, 2010 Tracking Survey. N=2,252 adults 18 and older; n=772 based on those who use a cell phone to access the internet.

Figure 7
According to Pew, more than half of cell phone Internet users go online daily from their mobile device.

	All adults	White, non-Hispanic	Black, non-Hispanic	Hispanic (English-speaking)
Own a cell phone	82%	85%	37%*	87%*
% of cell owners within each group who do the following on their phones				
Take a picture	76	75	76	83*
Send/receive text messages	72	68	79*	83*
Access the internet	38	33	46*	51*
Send/receive email	34	30	41*	47*
Play a game	34	29	51*	46*
Record a video	34	29	48*	45*
Play music	33	26	52*	49*
Send/receive instant messages	30	23	44*	49*
Use a social networking site	23	19	33*	36*
Watch a video	20	15	27*	33*
Post a photo or video online	15	13	20*	25*
Purchase a product	11	10	13	18
Use a status update service	10	8	13	15
Mean number of cell activities	4.3	3.8	5.4	5.8

Source: Pew Research Center's Internet & American Life Project, April 29-May 30, 2010 Tracking Survey. N=2,252 adults 18 and older, including 1,917 cell phone users. * = statistically significant difference compared with whites.

Figure 8
Demographic data on mobile device usage.

The Changing Face of the Digital Divide

Libraries have long been at the forefront of advocacy for increased broadband Internet access, particularly for the poor and for rural Americans. Mobile may not yet be the ideal solution for rural users, but the demographics of mobile Internet usage show encouraging signs of increased access among groups that have long been underrepresented among Internet users in the United States.

The report *Mobile Access 2010* from the Pew Internet and American Life Project notes that Latino and Black adults are more likely both to own mobile devices and to use them to access the Internet than their White peers (figure 8). While 80 percent of white adults own mobile phones, among African Americans and English-speaking Latinos the rate of ownership is 87 percent. Of all American adults with cell phones, 38 percent use them to access the Internet, but black and English-speaking Hispanic users far outstrip the average, at 46 percent and 51 percent respectively. Pew's survey did not have a Spanish-language option, so data on Spanish-speaking Latinos was not available.[21]

The Day for Mobile Services Has Come

The evidence is compelling. The vast majority of Americans now own cell phones. Nearly half use them to access the Internet. Sales of smartphones have already or soon will surpass those of traditional PCs. Underrepresented groups are accessing the mobile Internet in impressive numbers. Google is developing for mobile first and the desktop second. Apple is in the midst of making its desktop computers behave more like its mobile devices. If your library, like mine (and every library I can think of), has been transformed by desktop computing and Internet access, now is the time to take action and be proactive in providing robust services to mobile users.

Notes

1. Federal Communications Commission, Annual Report and Analysis of Competitive Market Conditions with Respect to Mobile Wireless, Including Mobile Services, 14th report (FCC 10-81), (Washington, DC: FCC, May 20, 2010), 92, http://hraunfoss.fcc.gov/edocs_public/attachmatch/FCC-10-81A1.pdf (accessed Jan. 4, 2011).
2. Amanda Lenhart, Cell Phones and American Adults (Washington, DC: Pew Research Center, Sept. 2, 2010), 2, http://pewinternet.org/Reports/2010/Cell-Phones-and-American-Adults.aspx (accessed Jan. 4, 2011).
3. Aaron Smith, Mobile Access 2010 (Washington, DC: Pew Research Center, July 7, 2010), 5, www.pewinternet.org/Reports/2010/Mobile-Access-2010.aspx (accessed Jan. 4, 2011).
4. Robert B. Kvavik, Judith B. Caruso, and Glenda Morgan, ECAR Study of Students and Information Technology, 2004: Convenience, Connection, and Control (Research Study 5, 2004), (Boulder, CO: Educause Center for Applied Research, 2004), cited in Shannon D. Smith and Judith B. Caruso, "Technology Adoption and Ownership of IT," chapter 4 of ECAR Study of Undergraduate Students and Information Technology,

2010 (Research Study 6, 2010), (Boulder, CO: Educause Center for Applied Research), 41, http://net.educause.edu/ir/library/pdf/ers1006/rs/ers10064.pdf (accessed Jan. 4, 2011).

5. Smith and Caruso, "Technology Adoption," 41.

6. Mike Abramsky, quoted in Philip Elmer-DeWitt, "Abramsky: Apple, RIM Could Triple Revenues by 2012," Fortune magazine website, Apple 2.0 blog, Aug. 18, 2009, http://tech.fortune.cnn.com/2009/08/18/abramsky-apple-rim-could-triple-revenues-by-2012 (accessed Jan. 4, 2011).

7. FCC, Annual Report, 92.

8. Federal Communications Commission, Annual Report and Analysis of Competitive Market Conditions with Respect to Mobile Wireless, Including Mobile Services, 14th report (FCC 10-81), (Washington, DC: FCC, May 20, 2010), 92, http://hraunfoss.fcc.gov/edocs_public/attachmatch/FCC-10-81A1.pdf (accessed Jan. 4, 2011).

9. Ibid., 16–17.

10. Ibid., 92.

11. Kristen Purcell, Roger Entner, and Nichole Henderson, The Rise of Apps Culture (Washington, DC: Pew Research Center, Sept. 15, 2010), 2, www.pewinternet.org/Reports/2010/The-Rise-of-Apps-Culture.aspx (accessed Jan. 4, 2011).

12. Smith and Caruso, "Technology Adoption," 41.

13. Smith, Mobile Access 2010, 2.

14. Ibid., 5.

15. comScore, "comScore Reports March 2010 U.S. Mobile Subscriber Market Share" (press release, May 6, 2010, http://comscore.com/Press_Events/Press_Releases/2010/5/comScore_Reports_March_2010_U.S._Mobile_Subscriber_Market_Share (accessed Jan. 4, 2011).

16. Smith, Mobile Access 2010, 17.

17. Smith and Caruso, "Technology Adoption," 47

18. Luke Wroblewski, "Mobile First," May 28, 2010, YouTube video on LinkedIn Tech Talks channel, www.youtube.com/watch?v=NjE_Or4VIlU (accessed Jan. 4, 2011).

19. Eric Schmidt, quoted in Nathan Eddy, "Google CEO Eric Schmidt at MWC Puts Mobile First," Feb. 16, 2010, eWeek.com website, Midmarket News, www.eweek.com/c/a/Midmarket/Google-CEO-Eric-Schmidt-at-MWC-Mobile-First-694942 (accessed Jan. 4, 2011).

20. Eric Schmidt, quoted in Chloe Albanesius, "Google's New Rule: Mobile First," Feb. 16, 2010, PCMag.com website, www.pcmag.com/article2/0,2817,2359752,00.asp (accessed Jan. 4, 2011).

21. "Back to the Mac," Apple Special Event, Oct. 20, 2010, www.apple.com/apple-events/october-2010 (accessed Jan. 4, 2011).

22. Smith, Mobile Access 2010, 4.

Chapter 2

Mobile Devices in 2011

Abstract

The exploding popularity of mobile devices has led to a market where new devices are being released, new programs are being developed, and new features are gaining and losing popularity very rapidly. This chapter of Libraries and Mobile Services *helps to make sense of the current market in mobile devices while explaining some of the basic principles of how these devices work and are differentiated.*

M any of us in the library world have been monitoring developments in mobile technology for some time. We've recognized that these devices offered new and compelling means for access to information and have long sought to take advantage of them with, let's be honest, limited success. You could be forgiven for feeling that mobile technology for libraries is old news. You might be concerned that despite the dramatic shifts in device ownership and usage outlined in the previous chapter, our responses are limited to tools like SMS messaging.

Thankfully, the shifts in adoption and usage have been mirrored by dramatic evolution in mobile devices themselves. I agreed to write this issue of *Library Technology Reports* in September 2009. I'm completing it in late 2010 for publication in early 2011. The eighteen months that have passed between when I undertook this project and its publication are a lifetime in the mobile world (in the case of a few mobile companies, quite literally a lifetime).

In September 2009, Android released the software development kit (SDK) for its 1.6 update. Since then there have been no fewer than five more new SDKs released, and many Android phones are now running OS 2.2. The Palm Pre, running the proprietary webOS operating system, had been on the market just a few months, and the GSM variant of the phone, which allowed it to run on AT&T's network, had just been released. Palm has since released two new versions of its phone, the Pre Plus and the Pixi. In 2010 the company was bought by Hewlett Packard, and in October, HP announced the Pre 2 and webOS 2.0. In the summer of 2009 Apple released the iPhone 3GS and celebrated the first birthday of the iPhone App Store, announcing that over 65,000 applications were available in the store and over 1.5 billion had been downloaded.[1] Two months later those numbers were 85,000 and two billion, respectively.[2] The most recent information available shows that by September 2010, the App Store had over 250,000 applications which had been downloaded over 6.5 billion times.[3]

If it seems that a lot has changed since I undertook this project, think of how different the mobile device landscape is from when Ellyssa Kroski wrote her LTR.[4] The devices available gratis with new mobile contracts have a hardware and software feature set the like of which was unthinkable in 2007. We used to talk about "smartphones" and "feature phones." Smartphones were those few devices that had PDA, e-mail, and rudimentary Web connectivity. Feature phones (sometimes referred to as "dumb phones") were those that were single-function devices, good primarily for making calls and texting. At best, feature phones sometimes had a very rudimentary WAP web browser, many of which were restricted to accessing a portal created by the wireless carrier, devoted to add-on services like paid ringtone downloads.

The era of the feature phone is at an end, and the dramatic difference between the devices Kroski addressed in her LTR and today's mobile devices can hardly be overstated. Today virtually every phone

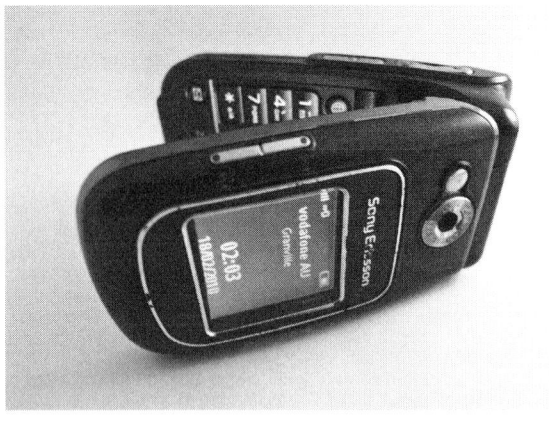

Figure 9
A "clamshell" flip phone (Photo credit: aeiss, "Phone," www
.flickr.com/photos/ohaithar/4365544298, licensed under the
Creative Commons Attribution 2.0 Generic license, http://
creativecommons.org/licenses/by/2.0/deed.en).

offered by the major U.S. wireless providers has a wide
range of hardware and software capabilities, including
unrestricted access to the open Web.

The opportunities for libraries have likewise
expanded. Improved screen technology allows better reading experiences. Third-party application markets are hotbeds of new information access models.
High-resolution cameras turn any mobile device into
a barcode scanner. GPS chipsets allow for innovative
location-based services.

The pages following are a crash course in the
mobile device features that have so profoundly
changed the capabilities of these devices, and by
extension, our relationships with them. These are the
hardware and software features that we need to keep
at the forefront of our service delivery strategies as we
enter the second decade of the millennium.

Hardware

Form Factors

Form factor refers to the hardware configuration of
a device, including its shape and available physical
methods of input. We've seen a number of trends in
mobile form factor over the years, such as the "flip
phones" many of us carried in the early to mid-2000s.

Flip phones, or "clamshell" devices (figure 9), are
those that fold in half, typically showing a small aux-
iliary display on the outside and concealing a larger,
higher-resolution screen and keypad within. If you
owned one of the approximately 50 million RAZR
phones sold by Motorola, you owned a flip phone.
Another early trend in mobile devices was the so-called
"candybar" form factor, which placed the keypad and
screen on the same side of a slab-shaped device. The
candybar seems like the inevitable product of the

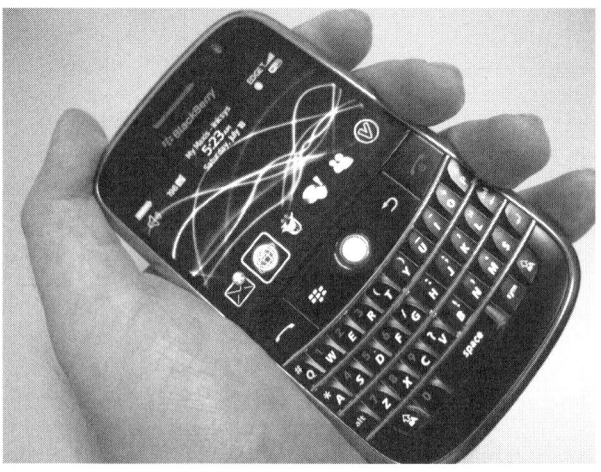

Figure 10
Phone with a QWERTY keyboard (Photo credit: Cheon
Fong Liew, "Blackberry Bold in hand," www.flickr.com/
photos/liewcf/3741159735, licensed under the Creative
Commons Attribution-ShareAlike 2.0 Generic license, http://
creativecommons.org/licenses/by-sa/2.0/deed.en).

ongoing miniaturization of the classic Zack Morris-
style cellular phone. If you ever owned a Nokia phone
that allowed you to play Snake on a black-and-white
screen, you owned a candybar phone.

With the notable exceptions of some wild experi-
ments on the part of Nokia and Samsung, the past few
years have seen a consolidation in form factor center-
ing on two key features: the QWERTY keyboard, and
the touchscreen.

QWERTY

Driven in large part by the innovations of RIM's
BlackBerry devices, various permutations of the full
computer keyboard on mobile devices (figure 10)
became quite popular, particularly in the corporate
enterprise world. These QWERTY devices (named after
the first six characters on a standard keyboard) wore
their productivity lineage on their sleeves. They were
optimized for rapid access to e-mail and instant messaging, and many an executive became as facile with
the BlackBerry keyboard as with the computer's. The
first few iterations of the BlackBerry didn't even have
phone capabilities and were strictly e-mail devices
that owed more to the alphanumeric pagers of the late
90s than to the mobile phones of the time. In many
ways, we've now come full circle, and the phone has
for many users ceased to be the central feature of their
mobile device.

Indoctrinated in many cases by their employers,
users soon began to demand the "push" e-mail and
productivity features of RIM's devices, and BlackBerry
eventually made a play in the consumer market. By
the late 00s, RIM's QWERTY devices were available
on nearly every carrier. These devices sat alongside

other QWERTY devices like Danger, Inc.'s Sidekick, an innovative device that was the first consumer-oriented phone with a true network focus, sporting an always-on data connection and excelling particularly in instant messaging capabilities. One of Danger's founders, Andy Rubin, later went on to create the Android mobile OS for Google.

Today there are many QWERTY devices on the market. Some retain the traditional BlackBerry form factor, with a large keyboard sitting below a squarish screen. Others conceal the QWERTY keyboard behind a large screen, allowing for basic functions when closed and sliding open when it's time to do some serious typing. However, even BlackBerry has been experimenting with form factors that eliminate the keypad entirely. The form factor that seems to rule the day is the slab-like touchscreen.

Touchscreen

The flagship touchscreen device of the past few years has been Apple's iPhone (figure 11). This is the device that proved that less can be more when it comes to buttons. There was great skepticism at the iPhone's launch about its touchscreen interface, particularly the on-screen keyboard. Some users accustomed to clicky keyboards, scroll wheels, and trackballs felt there was no way they could maintain the same comfort and efficiency when entering information on a device that provided no physical feedback.

What many users found, however, was that the keyboard was pretty decent, and the autocorrect software was great. Combined, the two served to make text input on the iPhone good enough to inspire sales of around 75 million devices before the end of 2010. More importantly, many users found that the direct manipulation of on-screen elements using the intuitive tap, swipe, and pinch gestures foundational to the iPhone's operating system was shockingly intuitive. Online videos have proliferated showing infants and toddlers presented with the iPhone for the first time, and invariably show the children swiping through photos or interacting with games with ease.

The usability of the iPhone's touchscreen owes much to the underlying software, but there's a hardware story here as well. The iPhone was one of the first consumer devices to use what's called a capacitive touchscreen, one that senses the presence of a finger by its effect on an electric field. This was a leap forward from the resistive touchscreens commonly used on PDAs and other devices before the iPhone. Resistive touchscreens rely on actual physical pressure to record input. The move to capacitive touchscreens allowed for more accurate input without the use of a stylus and for the simultaneous tracking of multiple inputs, so-called multitouch.

Touchscreens have become common on many

Figure 11
The iPhone (Photo credit: Yutaka Tsutano, "iPhone 4 32GB Black," www.flickr.com/photos/ivyfield/4731067532, licensed under the Creative Commons Attribution 2.0 Generic license, http://creativecommons.org/licenses/by/2.0/deed.en).

mobile devices. Google's Android OS is optimized for touchscreen devices, as is HP/Palm's webOS. As alluded to above, RIM has even released a series of touchscreen devices without its trademark keyboard, though the first few were hampered by resistive, rather than capacitive, screens.

Nokia, long the leader not only in phone sales, but in smartphone innovation, used to boast about the number of buttons on its high-end phones. Its phones had camera buttons, menu buttons, media playback buttons, number buttons, letter buttons, and many more. Some models topped out at more than 100 buttons, each dedicated to a single function. Compare this to the iPhone or Nexus One, Google's flagship Android device released in early 2010. Each has a number of buttons you can count on one hand. However, their touchscreens ensure that every application can implement as many custom controls as necessary, in effect an infinite number of buttons, customized for each application.

Components

Screens

It bears mentioning here that touch isn't the only evolution that's come to mobile device screens in the past few years. The phones available to most of us today when we initiate or renew a contract feature high-resolution color screens. These screens, as much as any other innovation, are what has enabled mobile devices to transcend simple "phonehood." Increasing the resolution of the display allows for more information per screen, more legible rendering of type, and quality display of photos that has likely relegated the wallet-sized print to the dustbin.

Screen resolution is described with two figures, pixel dimensions and pixel density. If you've ever fiddled around with hooking a laptop up to a projector, you're likely familiar with pixel dimensions. The projector in the conference room nearest my office maxes out at 800 × 600. These numbers refer to the number of individual pixels available on a screen to create an image. In the example above, the projector can use 600 horizontal lines of 800 pixels, each of which can be a single color at any given moment, to create the image on the screen. The pixel dimensions of mobile devices are rapidly approaching those of the computer monitors many of us used through the 2000s. One of the flagship Android devices of the summer of 2010, the HTC Evo, featured a 480 × 800 screen.

Pixel density takes both the pixel dimensions and the actual physical dimensions of the screen into account to provide pixels per inch, or PPI, a measure of a given screen's ability to display a smooth, detailed picture. Apple branded the screen used in the iPhone 4 as the "Retina Display" because the pixel density supposedly approaches the point where it is physically impossible for the human eye to distinguish individual pixels. This screen has pixel dimensions of 960 × 640 in a 3.5-inch display, for a pixel density of 326 PPI.

While you may still be able to find a phone featuring an old-fashioned black-and-gray LCD, the only monochrome devices I've seen on the market recently have been phones marketed to seniors, children, and others needing only a truly bare-bones device capable only of making phone calls, in some cases only to a predefined set of emergency numbers.

Cameras

Speaking of photos, many of the fascinating developments in mobile device software and interactions that we'll look at later in this report rely on the ever-improving cameras that are ubiquitous on today's phones. I'm not kidding about ubiquitous; it's come to the point that the absence of a camera is a feature for some enterprise mobile consumers. If you're the guy buying BlackBerrys for a Department of Defense contractor, you need to specifically order the model with no camera to keep the nation's secrets from flying out over the airwaves.

Describing camera quality is a challenge. The easiest measure is the camera's resolution, the number of pixels that the image sensor uses to capture the image. However, the hardware and software in digital cameras, including those embedded in mobile devices, have advanced to the point where raw megapixel count no longer suffices to quantify image quality. A five megapixel sensor can theoretically capture an image detailed enough that you can make photo-quality 8 × 10s, but if the camera's lens distorts, or the image capture software can't intelligently adjust exposure to

avoid shadowed faces or a blown-out white sky, those 8 × 10s won't find a place on your mantel.

That said, it's increasingly common for mobile devices to ship with embedded cameras that are objectively quite good. A couple of years ago, I spent many hours researching before buying a digital camera that I felt hit the sweet spot between image quality, lens and optical quality, zoom, and compact form factor. That camera has gathered dust for months as my phone has become my exclusive camera. Actually, I did take my stand-alone camera on a recent trip to Mexico. I didn't want to risk my phone getting wet at the beach.

Wi-Fi

While mobile carriers have been eager to push smartphone devices that require additional data service plans, thankfully, many of the Android, iOS, Palm, and BlackBerry devices available today also come equipped with Wi-Fi radios. With Wi-Fi, users can connect to their home or other Internet connections, often at speeds much greater than what's available through their data plan. iPhone users are well acquainted with the necessity of Wi-Fi, as AT&T won't allow users to download applications or media files larger than 20 MB over their data connection. Users running into this limitation are notified, hypothetically, that downloading The Secret of Monkey Island requires connection to a Wi-Fi network or syncing to a computer.

In smartphones with robust third-party software development communities, such as Android and iOS, the Wi-Fi radios have become a workaround source of connectivity for applications that appear to compete directly with the business of the mobile carriers. On iOS, for example, Skype developed a native application that allows users to make free voice calls to any Skype account holder, and to any phone number for a very small per-minute tariff. The application was approved through Apple's App Store, but it can connect only over Wi-Fi, not using AT&T's network.

GPS

Add the standalone car-mounted GPS unit to the list of devices made obsolete by multifunction mobile devices (figure 12). Beginning in the mid-2000s, smartphones had very primitive geolocation applications that used the unique names of cell towers to pinpoint a user's location. Google, for instance, shipped a Google Maps application for BlackBerry that used this cell tower information to place the user on a map. The original iPhone shipped with a similar location service in the Maps application.

The capabilities and accuracy of mapping and navigation applications increased greatly with the inclusion of actual GPS chips in mobile devices. Current-generation devices use a sophisticated combination

Library Technology Reports   alatechsource.org   February/March 2011 of cell tower, Wi-Fi access point, and GPS satellite navigation to pinpoint a user's location with surprising rapidity and accuracy. An iPhone user can quickly perform Google Maps searches and get directions between locations optimized for driving, walking, or in many metropolitan areas, even public transit. Most current Android OS devices ship with true turn-by-turn navigation that can speak directions aloud.

The location services in mobile devices isn't limited to maps and navigation. Increasingly, users can grant permission to third-party applications to selectively access location. A Yelp.com application, for instance, can point out nearby restaurants. The World-Cat application can show you the library nearest to your current location. Various fitness applications allow users to track their cycling or running routes. More on applications later.

Bluetooth

I'm sure we all still shudder when we recall the plague that afflicted the nation's yuppies in the mid-2000s. Previously competent and successful legions of suited businessfolk roamed our city streets and airports yelling loudly and gesticulating wildly, each sprouting a similar unsightly blinking growth from one ear or the other. They were the Bluetoothed, determined to infect every public space with loud one-sided conversations with what we can only assume were demonic tormentors.

Thankfully, the worst of the epidemic seems to have passed, and we can now focus on rebuilding the reputation of the otherwise benign radio standard known as Bluetooth. While Bluetooth radios have been common on phones for a number of years now, only recently have handset manufacturers been building in support for some of the more interesting features of the technology. No longer is it simply a means for connecting monaural headsets. Users can now use stereo Bluetooth headsets to listen to music without tangled headphone wires, and connect Bluetooth keyboards to turn their mobile device into a full-fledged productivity workstation.

Flash Memory

Not to be confused with Adobe's Flash multimedia technology, flash memory refers to solid-state data storage technology. In a process analogous to Moore's Law, the price and capacity of flash memory have declined and increased respectively at an impressive rate. The first iPod to ship with solid-state flash memory instead of a traditional hard drive was 2005's iPod Nano, the cheapest version of which was $199 for 2GB of storage. Today's iPod Shuffle offers the same storage for $49.

I mention iPods not just because Apple is the world's largest consumer of flash memory, but because the gap between mobile phones and multimedia players has narrowed to near-invisibility in recent years, largely because of the increased storage capacity of mobile phones. Where once phones shipped with a few megabytes of storage, suitable only for a handful of low-res camera phone photos or $2.99 ringtones, now it's common for smartphones to carry up to 32 GB of storage. Many mobile devices also allow for expanded storage using SD cards. Apps, MP3s, videos, high-res photos, and videos all require ample storage space, and it's still easy to max out the available space on a device.

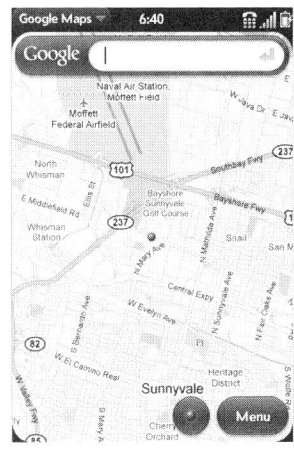

Figure 12
Using GPS on a webOS smartphone.

Batteries

Many of us have become accustomed to re-upping our phone contracts every couple of years and receiving a shiny new device in return for our continued loyalty. With each upgrade we've gotten a vastly expanded and improved feature set—Bluetooth, color screen, multi-megapixel camera, gigabytes of multimedia storage, video calling, and so on. There's been one notable exception to this otherwise uninterrupted march of forward progress: battery life.

Where once all but the heaviest users of mobile phones could expect to go days per charge, we've now entered an era where users of most devices of recent vintage need to charge them daily. It may be that battery technology hasn't quite kept pace with the other device innovations, but it is also true that all of the features of our new phones come with a price. Big bright color screens require far more energy to power than did the black-and-white LCDs of our Snake-playing Nokias. Likewise, Wi-Fi connectivity, Bluetooth, and GPS each rely on new power-hungry radios. Our devices have also become steadily smaller, allowing less space for batteries. There's also the inevitable consequence of increased functionality (see figure 13).

Software

Of course, what really differentiates today's smartphones from their predecessors isn't the hardware, which largely consists of predictable evolutionary

Figure 13
A valid point regarding battery life in today's devices (Source: rands, July 22, 2008, http://twitter.com/#!/rands/status/865671906 [accessed Jan. 4, 2011]).

steps from previous iterations, but the software, which has taken these devices into completely new realms of functionality. The software that's setting today's top-of-the-line phones apart from their ancestors comes in two broad categories: smartphone operating systems, which are developed or licensed by handset manufacturers, and applications that run within those operating systems, often developed by third parties.

Almost without exception, current-generation smartphone operating systems provide a suite of functions that until recently were available only on desktop and laptop computers, on specialized consumer electronics, or in PDAs that lacked always-on connectivity. Some of those features are as follows.

E-mail

Every smartphone operating system today ships with its own sophisticated e-mail client built in. The vast majority will support multiple e-mail accounts, allowing users to access, for instance, both a personal and a work e-mail account from the same device. Many of these e-mail clients provide native access to popular Web-based e-mail services like Gmail, Hotmail, and Yahoo! Mail. Increasingly, these mobile e-mail clients support Microsoft Exchange server. Exchange is perhaps the most common e-mail server for the enterprise, and Exchange support means many users can access their work e-mail securely, including server-side folders. Exchange support also allows server administrators to remotely wipe data from mobile devices securely.

Note: Most smartphone e-mail clients do support HTML e-mail messages, but designing HTML messages that degrade appropriately for the smaller screens and higher pixels-per-inch on mobile devices may be even more challenging than developing HTML for mobile websites. If you typically send users HTML e-mails, watch your usage statistics for mobile browsers and

user agents. If you see many users viewing your messages on mobile devices, consider reverting to plain text messages. If your marketing strategy requires the analytics provided by HTML e-mails, simplify your message layout to allow the e-mail client to zoom and reflow text for legibility.

Calendar

Exchange support also allows enterprise users to integrate Microsoft Outlook calendars with their mobile device calendars. The calendars built in to most mobile operating systems sync wirelessly with Exchange or CalDAV servers, allowing for real-time updates and notifications from services like Google Calendar or servers like Oracle Calendar (depending on your server configuration).

Multimedia

In many ways, today's smartphones are directly evolved from both past mobile phones and past portable multimedia players. BlackBerry smartphones have moved from business-oriented messaging, calendaring, e-mail, and phone capabilities into the consumer multimedia realm, adding support for MP3 and other audio and video file formats. Apple's iPhone evolved in large part out of the company's iPod line of media players, adding a phone and PDA functions to what was arguably the best media player on the market.

These converged devices offer increasingly robust support for multimedia. Many devices allow users to load media files by mounting the device on a PC like a USB drive. Users can then drag and drop files into the mobile device's onboard storage and watch movies or listen to music or audiobooks on the go. Apple's iPhone syncs to the iTunes desktop application to load media, rather than mounting as a drive. This approach has pluses and minuses. Users are unable to directly manipulate the files on their device, and moving a single file onto or off of the iPhone requires a full sync. The syncing process does offer the benefit of allowing users to specify, for instance, that episodes of podcasts that they've already listened to be removed from the device upon syncing, automatically freeing space for additional media. Windows Phone 7 syncs media using the Zune desktop client. A number of third parties offer desktop software that allows for a similar type of syncing for BlackBerry, Android, and Palm devices.

At launch, the Palm Pre purported to offer media syncing through iTunes, a feature that would have made the Pre unique among non-Apple devices. This syncing capability was achieved without Apple's permission, and subsequent versions of iTunes disabled the feature.

The system-level support for many multimedia

codecs means that not only can smartphone users load audio and video files from their personal libraries, but they can often access media files embedded in web-pages and otherwise available online.

Web Browsers

Perhaps the most important, and likely one of the more underrated aspects of current mobile device software is the mobile web browser. As more and more of the applications we rely on in our work and personal lives have moved to the cloud, the reliable, speedy, standards-based web browser has done more to legitimize mobile devices as true productivity tools than just about anything else.

Today all of the major smartphone operating systems ship with a web browser that rivals major desktop platforms in its support for web standards. In addition, these browsers offer important affordances for the smaller screen sizes of mobile devices. They allow for intelligent zooming of webpages, identifying sections of the page and zooming to fit columns of text, and so on.

The signature evolution of this generation of mobile devices, however, is likely the rise of widely available high-quality third-party software applications. This is the age of the app. These programs range from bite-sized to multi-gigabyte, and from games to medical and automobile diagnostics. They take the bare essentials of the mobile device—a network connection, a camera, geolocation, and so on—and using the highly configurable touchscreen interfaces on most devices, transform your phone into something entirely other. Today apps are available for many smartphone platforms, with varying degrees of quality. In this report I'll focus on five operating systems: iOS, Android, webOS, BlackBerry OS, and Windows Phone 7.

iOS

iOS, previously called iPhone OS, launched with the first iPhone in 2007. The operating system is derived from Apple's OS X operating system, which is used on its laptop and desktop computers, as well as Apple servers. The OS is optimized for the few pieces of hardware upon which it runs legitimately: the various models of the iPhone, iPod Touch, and iPad. As befits these devices, the user interface for iOS is touch-centric and app-centric (figure 14). Every function of the device is contained within an application. The Settings application contains general device settings and preferences.

The iPhone launched in 2007 with just a small handful of native applications, all of which were developed by Apple. In addition to the core Phone, Messages (for SMS and MMS text messaging), Mail, and iPod applications, the iPhone included the Calculator

Figure 14
The home screen on the iPhone 4. (Photo credit: William Hook, "iPhone 4 - Retina Display - Home Screen," www.flickr .com/photos/williamhook/4742863198, licensed under the Creative Commons Attribution-ShareAlike 2.0 Generic license, http://creativecommons.org/licenses/by-sa/2.0/deed.en).

app, Maps app, YouTube app, Notes, and so on. While many suspected that it might eventually be possible for third-party software developers to write native iPhone applications, initially Apple did not offer any means to do so, and instead touted the ability for developers to write robust web applications.

Unsatisfied with the limited capability of web applications, third-party developers soon discovered ways to break in to the iPhone's closed file system and load their own applications onto the device. The process of circumventing the security measures in the iPhone's software and hardware is known as "jailbreaking." The popular iPhone applications Twitterrific and Tap Tap Revenge began as samizdat software for jailbroken iPhones before it was possible to develop legitimately and before any documentation was available for developers. Late in 2007 Apple announced the forthcoming release of an official software development kit (SDK) for the iPhone, which would allow developers to create native iPhone applications and to submit them for sale through the forthcoming App Store.

Applications for iOS devices are written using a language called Objective C, the same language used for applications in Mac OS X. Prospective iOS application developers who register with Apple can download for free the software development kit (SDK) for developing iOS applications. The (hefty) download includes Apple's Xcode application, an integrated development environment used for creating both iOS and Mac OS X applications. The SDK also includes iPhone and iPad simulator applications, which allow developers to test their applications on screen. These simulators, like Xcode, are OS X-native applications, and will not run on Linux or Windows. Also included in the SDK is Dashcode, an application for developing iOS-optimized web applications.

iOS applications are available only through the iTunes App Store, either in the iTunes desktop application, available for Windows or Mac OS, or directly on iOS devices themselves. Dating back to the days before the App Store existed, there has always been an appetite among some developers and users to sidestep the App Store in order to install directly on the device software that wouldn't pass Apple's rigorous approval process. Installation of unapproved third-party software requires the process euphemistically referred to as jailbreaking. This process also allows users to bypass the so-called "carrier lock" that keeps the iOS device tied to AT&T's network, enabling its use on any compatible network.

Though jailbreaking has become quite popular as a way to customize one's iOS device or to run an iPhone on T-Mobile's GSM network rather than AT&T's, it is also a risky proposition. Jailbreaking relies on exploiting holes in iOS's security model, holes which could also be used to compromise users' personal data on the device. The jailbreaking community catalogs known exploits, but takes care to use only one at a time, knowing that Apple will always patch whatever vulnerability has been exposed by the latest jailbreaking method. This means that at any given time there are a number of outstanding exploits awaiting future use as jailbreaking methods that could in the meantime be put to nefarious use.

Android

The roots of the Android OS (figure 15) lie in the early development of another smartphone, the Sidekick. Created by a company called Danger, Inc., the Sidekick was notable as one of the first consumer-oriented device with a full QWERTY keyboard and with so-called "push" technology, which allowed for near-instantaneous delivery of data to the device. Previously available only to enterprise customers who could deploy a BlackBerry server, push data service provided not only e-mail delivery, but always-on instant messenger capability. The Sidekick's implementation of AOL's AIM instant message protocol made it popular not only with its target teen and twenty-something demographic, but with hearing-impaired users.

Under Danger, Inc.'s direction, the Sidekick went through a number of hardware revisions over several years on T-Mobile's network. In 2008, Danger was purchased by Microsoft. Shortly thereafter, Danger's former CEO, Andy Rubin left Microsoft to found a company called Android. Android was subsequently purchased by Google, which further developed and released its mobile operating system as Android OS. Andy Rubin has stayed on at Google, where he oversees Android development.

Free as in Android

Android is unique among the major smartphone OSes in that it is ostensibly free and open source, though in Android's case, both of these terms require some caveats. Beginning with open source, Android is not strictly a community-driven open source project like many of the standard bearers of free/libre/open source software (FLOSS). Key aspects of Android OS development are completely contained within Google. As stated on the Android Open Source Project page, "Some parts of Android are developed in the open, so that source code is always available. Other parts are developed first in a private tree, and that source code is released when the next platform version is ready."[5]

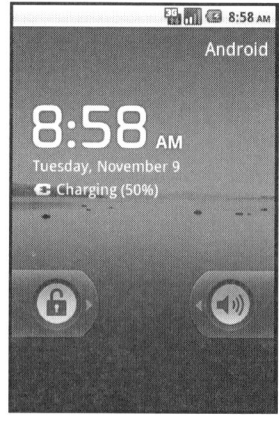

Figure 15
The Android operating system.

Google has released the source code of Android under the Apache software license. The Apache license differs from the GNU Public License (GPL), which uses a more traditional open source model and requires that derivative software also be released as open source. Under the Apache license, developers and handset manufacturers can add proprietary code to their implementations of Android without having to release it back to the open source community.

So, Android is open source and thus free as in freedom. For handset manufacturers and mobile carriers, it is also ostensibly free as in beer, but in practice that may not actually be the case. While Google will not charge anyone for Android, there are a number of companies that assert that Android implementations infringe on their intellectual property, and they have begun filing lawsuits demanding license fees.

One of the companies that is targeting Android devices with licensing suits is Microsoft, which has successfully extracted license fees from phone manufacturer HTC for its Android handsets.[6] HTC has long been a partner with Microsoft on Windows Mobile devices, and so the matter of patent licenses for HTC's Android phones was handled quietly and out of court, meaning we don't know much about what intellectual property was at stake or what fees HTC paid.

Microsoft CEO Steve Ballmer, seemingly very confident, asserted in a recent Wall Street Journal interview that implementing Android would not be free for handset manufacturers, "Android has a patent fee. It's not like Android's free. You do have to license patents. HTC's signed a license with us and you're going to see license fees clearly for Android."[7]

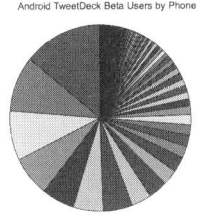

Figure 16
Android has become popular and its usage diverse (Source: Toby Padilla, "Android Ecosystem Infoporn Overload," TweetDeck blog, Oct. 12, 2010, http://blog.tweetdeck.com/android-ecosystem [accessed Jan. 4, 2011]).

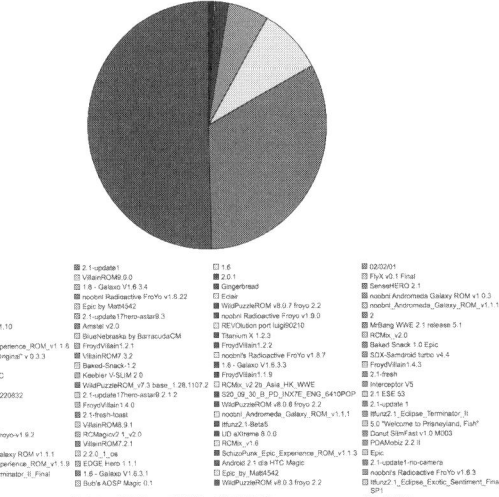

Figure 17
Another illustration of the diversity of Android usage.

Android Adoption, Implementation, Fragmentation

So far, the threat of patent license fees doesn't seem to be slowing the rate at which handset manufacturers are adopting Android and turning out handsets. Over the past year and a half or so, best-of-breed Android handsets have been released at a staggering rate (figure 16). Between the summer and fall of 2010, Android surpassed iOS in the number of phones shipped per month (note that iOS also ships on nonphone devices). A recent study released by the developer of the popular crossplatform Twitter client TweetDeck showed that among 36,427 Android users, there were nearly 250 distinct phone models in use.[8] Every major U.S. mobile carrier has at least one Android handset on offer, and many have a half-dozen or more.

While this explosion in Android adoption and handset models bodes well for the future of the platform and for prospects of development targeted at Android, it also speaks to some challenges. In order to provide the best possible user experience, app developers must take hardware into account as they develop software, and the sheer number and variety of Android handsets on the market makes this task very difficult.

Likewise, because of Android's liberal software license, it's not necessarily even appropriate to speak of it as a single platform. Handset manufacturers like HTC and Motorola, as well as carriers like Verizon, have been shipping handsets with versions of Android

so highly customized that they are incompatible with future updates to the Android OS, which instead have to be similarly customized before being shipped to users. The same survey by TweetDeck found over 100 different variations on the Android OS were running on their users' phones (figure 17).[9] While a significant majority were using recent mainline Android releases (2.2 and 2.1), a still-significant number were using outdated or splinter versions of the OS that would likely be incompatible with applications developed for the current Android release.

Android App Development

Android development is very similar to Java development—so similar, in fact, that Oracle, the owner of key Java patents, has sued Google for infringement. Regardless, developers who are accustomed to developing Java applications should be able to transition to Android applications with little difficulty. Google makes a number of tools available to developers in its Android Developers site. Here developers can contribute to the open portions of the Android source code, as well as read documentation and best practices for developing native Android apps. The Android SDK is also available and includes a desktop emulator developers can use to test applications. Google's recommended method for writing applications is to use the crossplatform Java-based Eclipse IDE, for which it provides an Android Development Tools plugin.

Android Developers website
http://developer.android.com

Developers wishing to publish or sell their Android application through Google's Android Marketplace can pay a $25 fee to register as a Marketplace Developer, after which they can submit applications. The aforementioned hardware and software fragmentation in Android extends to app stores, however. First, not all Android devices are compatible with the Android Marketplace. Google limits Marketplace access to devices that meet a stricter set of guidelines, among which at the time of writing was the need for a persistent network connection from a mobile carrier. This requirement has so far hampered the emergence of a contract-free iPod Touch equivalent running Android. Google's Android Marketplace is not the only game in town, however. Both Verizon and Amazon.com have announced intentions to open Android app stores, and it's speculated that handset manufacturer HTC will do the same.

webOS

The launch of the Palm Pre (figure 18) at the 2009 Consumer Electronics Show was seen as a welcome return to form for the company that defined the PDA in the 90s and pioneered the PDA/phone hybrid in the 00s with the Treo. Industry pundits, watching Palm's balance sheets carefully, predicted that the Pre was a do-or-die move for the company. In the years leading up to the Pre's launch, Palm had become a haven for a number of ex–Apple employees, including Jon Rubinstein, who lead Apple's iPod division. It came as little surprise then, that the Pre featured both hardware and software worthy of an Apple device, polished, thoughtful, and innovative.

The Pre featured a multitouch touchscreen interface, the first commercially available phone other than the iPhone to have one, a feature that provoked talk of lawsuits from Apple. Unlike the iPhone, it also featured a slide-out hardware keyboard and a small trackball, both of which were entirely optional, as all device features could be operated with the touchscreen. Palm later introduced the Pixi, a slimmer version of the Pre that sported a fixed keyboard below the screen. The Pre was the first smartphone to cleverly integrate a number of Web-based services, allowing users to sign in to Facebook, Twitter, and Gmail, and then see contacts from all three services seamlessly integrated into the phone's contact list. The Pre also wisely included an application that emulated the earlier Palm OS, allowing longtime Palm users to load dearly loved applications that no longer run on any other device.

The operating system running both the Pre and Pixi is known as webOS, an appropriate name since even native applications for the device are largely written using web standards and programming languages. webOS applications can be written entirely in HTML, CSS, and JavaScript, using the webOS Mojo

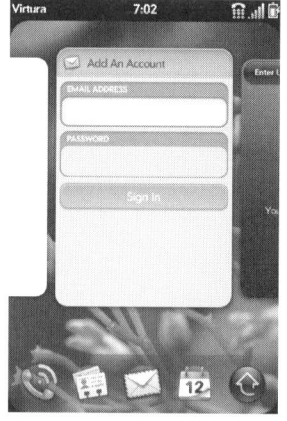

Figure 18
webOS on the Palm Pre.

JavaScript framework, which provides access to the native device functions. Applications can be submitted for distribution through the webOS App Catalog, where they are available on users' devices or through RSS feeds.

The barrier to entry for writing native webOS applications is by far the lowest of all the major smartphone platforms. Yet, a recent survey of application developers by Millennial Media, a mobile advertising firm, found that only 4 percent intend to target webOS in 2011.[10]

The Pre may not have been a do-or-die device for Palm. While it was generally a critical success, it saw only modest consumer adoption, perhaps in part because it launched exclusively on Sprint's network, one of the smallest of U.S. carriers. Just over a year after the launch of the Pre, Palm was acquired by Hewlett-Packard, which now markets the Pre as the HP Palm Pre, and webOS as HP webOS. There is every indication that HP intends to throw considerable corporate weight behind the development of webOS devices. webOS 2.0 was launched in late 2010 alongside the Pre 2. Industry watchers speculate that HP's interest in Palm and webOS had as much to do with future developments of tablet computers as with the smartphone business.

BlackBerry

If there is a granddaddy of mobile information access, it's the BlackBerry. Developed by Canada's Research in Motion in the late '90s as a device that had more in common with a pager than a phone, the BlackBerry was for many years the only option for users who needed access to e-mail at all conceivable times. The BlackBerry has been synonymous with harried executives and thumb-cramped attorneys for many years, and only in the past several years has it made inroads into the consumer market.

The first BlackBerry devices were small, with a monochrome screen allowing users to read or compose just a few lines of a message at a time. There was no phone, no calendar, no games, just e-mail. But what e-mail! Using a BlackBerry required installation of the BlackBerry Enterprise Server on a company's e-mail server. Once it was installed, messages would be delivered to BlackBerry devices in real time, allowing for near-instantaneous e-mail communication anywhere. The device feature set expanded,

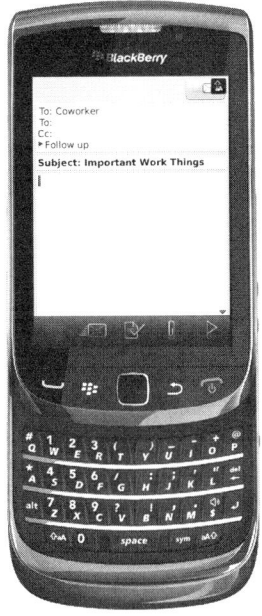

Figure 19
Using e-mail with a QWERTY keyboard on a BlackBerry device.

eventually incorporating all the features one would expect from a phone and PDA. By the mid-00s, RIM developed the BlackBerry Internet Service, which allowed cellular providers to run a BlackBerry server and offer to consumers the same lightning-speed e-mail delivery previously available only to corporate clients.

BlackBerry's run at the consumer market was remarkably successful. Users who'd become accustomed to the Black-Berry service with phones issued by their employers snapped up BlackBerrys for personal use, and devices like the Pearl, with a more pocketable form factor than the traditional BlackBerry slab, sold quite well. Unfortunately, while BlackBerrys excel in e-mail and instant message delivery, they have largely failed to offer quality Web access, using an antiquated and substandard browser. Likewise, the BlackBerry interface, no doubt a model of efficiency, has lacked the polish and design of competing products. Whether for these or other reasons, BlackBerry sales have suffered as Android and iPhone sales have soared, particularly in the enterprise.

RIM has made steps to compete, some more successful than others. In response to the iPhone, RIM released its first touchscreen BlackBerry, the first BlackBerry without a physical keyboard. Unfortunately, the device was roundly panned for having an unresponsive touchscreen that inhibited the extremely rapid input that BlackBerry users craved. Subsequent iterations of BlackBerry touchscreens have been better received. Likewise, with the release of BlackBerry OS 6.0 in late 2010, RIM has implemented a new web browser based on the industry standard WebKit, allowing for web rendering parity with other platforms.

BlackBerry offers a great deal of documentation for potential developers on its developer portal. Native BlackBerry application development is accomplished using Java, and the only supported method is to use the Eclipse IDE on a Windows machine. The BlackBerry SDK also provides a Windows-based emulator, allowing developers to preview applications on their desktop machines. Developers can submit applications to BlackBerry App World, a marketplace for

Figure 20
Social media applications on the touchscreen of a BlackBerry device.

BlackBerry applications that is accessible both on Black-Berry devices and via the Web.

BlackBerry Developer Zone
http://us.blackberry.com/developers

BlackBerry App World
http://appworld.blackberry.com/webstore

Windows Phone 7

Microsoft's Windows has quite a legacy in the smartphone world, and yet is also a complete newcomer. After years of development on the Windows Mobile platform, in 2010 Microsoft launched Windows Phone 7 (figure 21), which represented a complete clean break from its earlier mobile operating system. The business model for Windows Phone 7 is strictly as a software platform, much like the Windows desktop OS. Microsoft does not manufacture any of the Windows Phone 7 handsets, but instead licenses its operating system to manufacturers like Dell and HTC, which then build phones according to a fairly strict set of specifications determined by Microsoft. The handset manufacturers then sell their phones to mobile carriers.

The initial rollout of Windows Phone 7 devices was fairly well received by tech reviewers and consumers, though it remains to be seen if it will gain a foothold in the already crowded smartphone OS market. Like Android and BlackBerry, Windows Phone 7 had the advantage of having handsets available on multiple carriers at launch. In many ways, Windows Phone 7 devices mirror the first-generation iPhones: they lack meaningful multitasking and copy-and-paste clipboard functionality. However, they do have access to a third-party app marketplace at launch.

Additionally, Windows Phone 7 devices feature robust integration with other Microsoft tools and

Figure 21
Windows Phone, one of the newest mobile devices (Photo credit: Luca Viscardi, "windows phone 7 - 1," www.flickr .com/photos/viskas/5104024799, licensed under the Creative Commons Attribution-ShareAlike 2.0 Generic license, http:// creativecommons.org/licenses/by-sa/2.0/deed.en).

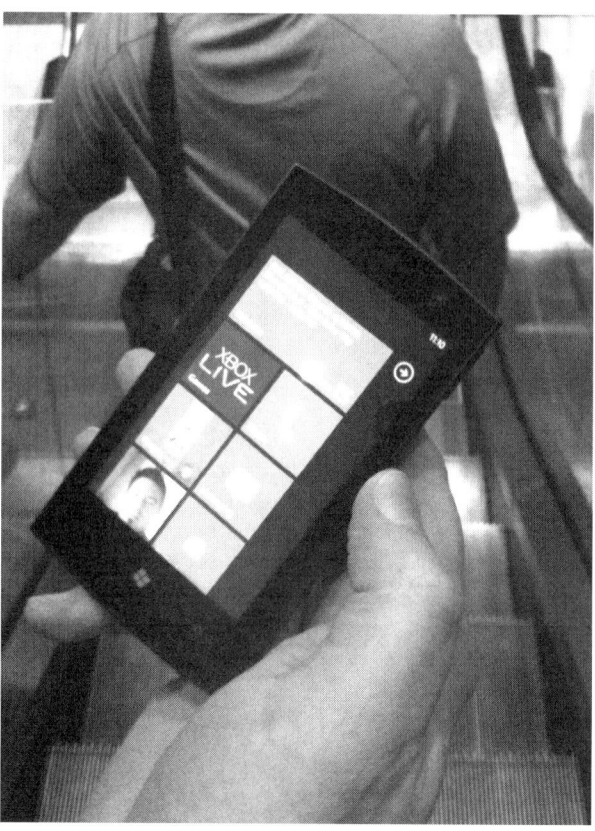

Figure 22
Gaming on a Windows 7 phone (Photo credit: Jeff Wilcox, "Windows Phone 7 Series," www.flickr.com/photos/jeffwilcox/ 4435367457, licensed under the Creative Commons Attribution 2.0 Generic license, http://creativecommons.org/ licenses/by/2.0/deed.en).

services. Each handset ships with a native Office suite that integrates deeply into the OS, allowing users to open and edit Office documents from their e-mail or from the Web. Users of Microsoft's Zune subscription service will be able to access their music and videos seamlessly on the device either by syncing files from a PC or Mac or by streaming. The device also integrates with the Xbox LIVE online gaming service, providing access to users' Xbox friends list and achievement points, which can also be earned by playing games on Windows Phone 7 (figure 22).

The stakes are high for Microsoft. Every indication is that mobile devices are poised to be as revolutionary as were desktop PCs. Microsoft has prospered for the past 20-odd years based on its dominating position in the PC software marketplace. You can argue that mobile is still in early days, but Microsoft is to date playing also-ran. It's also had some embarrassing missteps. Microsoft's tardiness in developing a state-of-the-art smartphone OS saw its most faithful handset partner, HTC, move strongly into the development of Android phones. An earlier attempt at a consumer-friendly, social-networking-focused phone, the Kin, was a dramatic failure. Released to great marketing fanfare, the Kin One and Kin Two were abject sales failures, prompting NYTimes.com to title a page on its website "A Youthful Market Spurns the Wares of Microsoft."[11] Only six weeks after the Kin launched, Microsoft killed the product altogether.

Windows Phone 7 is in its infancy, but there is every indication that Microsoft is making a serious run at the consumer smartphone space with these devices, and that it's doing things right so far. You could argue that Microsoft is late to the app-enabled smartphone OS party, but seemingly it has taken the past couple of years to learn from some of its peers' mistakes.

Unlike the BlackBerry or webOS app stores, which have a very limited selection of available applications, or the Android Marketplace, which can seem overrun with worthless or even scam applications, Microsoft has taken steps to ensure that its store is front-loaded with quality applications. Despite the fact that Apple's hands-on and sometimes overly restrictive App Store approval process raised the ire of some developers, Microsoft has instituted a very similar process for its store. Additionally, Microsoft reportedly approached the developers of a number of popular iOS and Android applications with very generous bonuses for developing Windows Phone 7 applications for launch.

Notes

1. Apple, "Apple's App Store Downloads Top 1.5 Billion in First Year" (press release), July 14, 2009, Apple Press Release Library, www.apple.com/pr/ library/2009/07/14apps.html (accessed Jan. 5, 2011).

2. Apple, "Apple's App Store Downloads Top Two Billion" (press release), Sept. 28, 2009, Apple Press Release Library, www.apple.com/pr/library/2009/09/28appstore.html (accessed Jan. 5, 2011).

3. Apple, "Statement by Apple on App Store Review Guidelines" (press release), Sept. 9, 2010, Apple Press Release Library, www.apple.com/pr/library/2010/09/09statement.html (accessed Jan. 5, 2011).

4. Ellyssa Kroski, "On the Move with the Mobile Web: Libraries and Mobile Technologies," *Library Technology Reports* 44, no. 5 (July 2008).

5. "Frequently Asked Questions," Android Open Source Project website, http://source.android.com/faqs.html (accessed Jan. 4, 2011).

6. Microsoft News Center, "Microsoft Announces Patent Agreement with HTC" (press release), April 27, 2010, www.microsoft.com/Presspass/press/2010/apr10/04-27MSHTCPR.mspx (accessed Jan. 4, 2011).

7. Steve Ballmer, quoted in Nick Wingfield, "Ballmer Aims to Overcome Mobile Missteps," Oct. 3, 2010, WSJ.com, http://online.wsj.com/article/SB10001424052748703466104575529861668829040.html (accessed Jan. 4, 2011).

8. Toby Padilla, "Android Ecosystem Infoporn Overload," TweetDeck blog, Oct. 12, 2010, http://blog.tweetdeck.com/android-ecosystem (accessed Jan. 4, 2011).

9. Ibid.

10. Millennial Media, State of the Apps Industry Snapshot (Baltimore, MD: Millennial Media, Oct. 2010), 2, www.millennialmedia.com/research/stateoftheapps (accessed Jan. 4, 20110.

11. Ashlee Vance, "Microsoft Calling. Anyone There?" July 4, 2010, NYTimes.com, www.nytimes.com/2010/07/05/technology/05soft.html (accessed Jan. 4, 2011).

Chapter 3

Mobile Solutions for Your Library

Abstract

This chapter of Libraries and Mobile Services *puts the information from the previous chapters into a library-specific context. By examining what different libraries have already done to provide mobile services and providing best practices and suggestions for future implementation, librarians gain a foundation for implementing or expanding services in their own facilities.*

So, if the preceding chapters have been convincing, and you now understand that your library needs to begin strategizing for mobile access and delivery of services to mobile users, how do you start?

Become a Mobile-Only User

If you already have a mobile device of recent vintage, commit to using it as your primary device for accessing your library's site and systems. If you are an employee of an academic or special library or other library where you are not also a frequent patron, consider becoming a mobile-only user of your local public library.

Give it a week or two. Document the experience. Use your website, OPAC, licensed resources, reservation forms, chat service, guides, the works. Don't limit yourself to those systems created by your library. If your users access it through your website, you bear some responsibility for the user experience, whether it's within your control or not.

Keep track of those resources or systems that automatically detect your mobile device and either notify you of a mobile interface or direct you there automatically. You will likely find some instances where you hit dead ends or where you're unable to access files, pages, or systems. Note especially when user-facing interfaces or systems are challenging or impossible to navigate. Are there interfaces in which you consistently found yourself inadvertently hitting the wrong links or buttons? Systems that appear to serve up files that your device is unable to load? How much longer does it take you to complete tasks as compared to your usual workflow?

As discussed above, e-book readers are available for nearly every smartphone platform, and

Figure 23
The University of Minnesota Libraries' website viewed in Mobile Safari.

most of the applications are free to install. Commit to using one of these tools to read your next book. Choose one of the many free public-domain titles that are available in these applications. Or take the plunge in the name of research and buy an e-book from the Kindle, iBooks, or Nook store. If your library subscribes to OverDrive (figure 24) or another e-book platform, try it, not as a library staff member, but as a reader.

This may be a humbling or frustrating experience. It will be eye-opening. It may also have very positive aspects. Note if there are situations where your mobility allows you to accomplish tasks that you couldn't if you needed to be seated at your regular workstation.

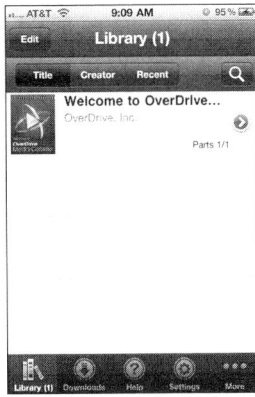

Figure 24
OverDrive, an e-book app popular with libraries.

Did you see another passenger on the bus or train reading a book that you'd been meaning to request? Were you able to request it at that very moment from your mobile device rather than hoping to remember once you got to your desk? Were you able to read on your mobile device long into the night without disturbing your partner with a bedside light?

Keep track of all your observations, both positive and negative, and share them with your colleagues. Encourage them to do the same. You will quickly be able to develop a list of areas where you should improve your mobile user experience.

Device Focus Group

Very few of us own or carry more than one current-generation mobile device, even gadget obsessives like your author (though once there's a contract-free Android equivalent of the iPod Touch, I expect that to change). The mobile device landscape is diverse, and becoming more so each day as new and worthy competitors to the established platforms enter the stage. So while our individual experiences with a single mobile device are instructive in creating a mobile strategy, they will not represent the diversity of our user base.

Canvass your staff, users, friends, family, whatever it takes, and find a willing group of people who carry a representative sample of the most popular smartphones and operating systems on the market. If I were convening such a group today, I'd hope to have a user each with an iPhone, a Palm Pre or Pixi, a device running the current and at least one past version of Android (such as 2.1 and 2.2), and user each for a recent BlackBerry device running the 6.0 version of the operating system and one with an earlier version. Windows Phone 7 devices may take some time to penetrate the market, but if you can find someone who uses one, so much the better. If possible, convene the users together on-site with a few of your public services, systems, and IT staff. It may be useful to use an opaque projector to display users' handsets on a screen so not everyone has to hover over the users' shoulders. Appoint at least one person to take detailed notes.

Ideally your focus group would consist of more users than staff, in which case you can begin by asking about their typical usage patterns for your services and systems. Do they typically access your systems from their home or work? Or from workstations in the library? Do they use your licensed databases, indexes, or e-books? If so, which ones? Do they often reserve books or other materials online for pickup in the library? Perhaps most importantly, are they already using their mobile device to interact with the library? If so, how?

Once you've established a list of your users' typical tasks, ask each user to walk your staff through any library tasks for which he or she is already typically using a mobile device. As is often the case, expect to see some usage patterns that you don't expect. Rarely do our preconceived notions of our users' methods match the reality. Next, ask users to step through typical nonmobile library tasks. Can they easily browse your collection? Access their account information?

With each of these sites, services, and tools, document what works, how well it works, and what doesn't work. You'll likely find that with ample zooming, tapping, and fiddling with tiny controls, your mobile users can accomplish a great deal with their devices. You will also likely find some tasks they can't accomplish. Finally, expect to find some tasks that, though possible, are challenging or time-consuming enough in the mobile environment that your users are unlikely to attempt them again.

This process should give you a good idea of where your library currently stands: which of your tools and services are usable, which may be frustrating, and which are downright broken for mobile users on various platforms. Once you have this list to start from, you can undertake one of four methods for beginning to develop your mobile service:

- Revamp your website for mobile-friendliness.
- Develop a mobile website.
- Develop mobile web applications (by which I mean applications that run within a mobile device's web browser).
- Develop native mobile applications.

We'll explore these options in the next sections.

Ensure Mobile-Friendliness in Your Current Site

In your audit of your website and other tools using various mobile devices, you likely found that the vast majority of tasks can be accomplished by mobile users, provided they have requisite patience. You also likely found some situations where users are stymied by code or multimedia that was designed for desktop users without regard for alternative platforms. Even if you have limited staff resources, there are some steps you can take to improve your existing site and make it more mobile-friendly.

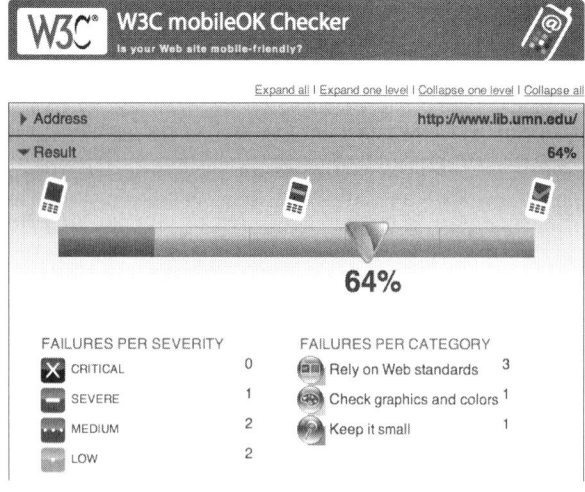

Figure 25
The W3C mobileOK Checker can help you verify the functionality of your application.

Validate

First things first. Even though you have anecdotal evidence of what works and what doesn't, you ought to run your site through the World Wide Web Consortium (W3C) mobileOK Checker (figure 25). This Web-based tool will evaluate the extent to which your website adheres to the W3C's Mobile Web Best Practices, a set of standards for mobile-friendliness of web documents.

W3C mobileOK Checker
http://validator.w3.org/mobile

W3C's Mobile Web Best Practices 1.0
www.w3.org/TR/mobile-bp

If your site is like most, the mobileOK Checker will provide you with a list of "failures" ranked by severity (Critical, Severe, Medium, and Low) and category. The failure categories correspond directly to specific items in the Mobile Web Best Practices document, and links to the relevant best practices are provided in line with the errors. Failure categories include "Rely on web standards," "Keep it small" (which refers to the total file size of your site), and "Stay away from known hazards," among others.

You will likely find that some of the failures reported by the mobileOK Checker are easily fixed by updating your page headers to properly reflect your site's character encoding and doctype. Sadly, while it's good to get these cleared up in order to avoid throwing errors in users' browsers, doctype and character encoding are unlikely to be the cause of your site failing for mobile users.

Fixes for Mobile-Friendliness

Give Your Site Some Space

I wear size XL gloves, XXL in some brands. Despite my gargantuan fingers, I've become quite facile with the virtual keyboards shipping on today's mobile devices, largely, I expect, due to the very clever error-correction software onboard in mobile operating systems. I'm not so lucky when it comes to using the Web on mobile devices. When I use sites that are not designed for mobile, I typically find myself tapping, pinching, zooming, and ultimately using just the very tip of my finger to hit links and buttons that were clearly designed for mouse and keyboard or for elven fingers. Here's a case where mobile-friendly design is probably just good design. Pad your navigation elements. Limit the number and proximity of links within text. Simplify.

Avoid Mouseovers

No mouse, no mouseovers. Makes sense, right? Some mobile web browsers will adjust by interpreting a tap on an element as a mouseover when one is specified in the code. At best, this means that flyout menus require twice as many taps for mobile users as they would for mouse users: tap once to expand the menu, tap again to select an element or expand a submenu (figure 26).

In the case of the menu in figure 26, selecting Bottle Cappers on an iPhone would require tapping once on Equipment to expand the Equipment menu, once on Bottling to expand the Bottling submenu, once on Bottle Cappers to invoke the mouseover change to the darker color, and then once more to follow the link. Or so I hear. The browser in webOS interprets a tap only as a click, meaning that it's impossible to expand the flyout menus. Tap on Equipment in your webOS browser, and you'll go to the full Equipment page, though you may see the menu flash on the screen briefly. Android's browser splits the difference, and a tap on Equipment both follows the link to the Equipment page and also opens the Equipment menu, with no clear means to close it.

Of course, these flyout menus are ubiquitous for a reason: they're a really easy way to provide quick access deep into your site. If you feel your site's navigation requires this kind of nesting within a page, try not to double up on functions within a single button. Use a button or label to expand a menu or as a link to a page, not both.

Mobile-Friendly Multimedia

A note to sensitive fanboys and fangirls: I'm about to use the F word. Flash. Much has been made of its presence or lack thereof on mobile devices. At present,

Library Technology Reports   alatechsource.org   **February/March 2011**

Apple has shipped tens of millions of mobile devices that do not, and likely will not ever, run Flash. Even Android doesn't ship with Flash enabled by default. Avoid Flash.

Fortunately, the vast majority of online video services are seamlessly serving up non-Flash versions of their content to mobile devices and other Flash-free platforms. This means that if your library's tutorial videos or screencasts are already up on YouTube, Vimeo, or another major video provider, you may not need to worry about converting them.

If you are encoding and hosting your own Flash video, consider re-encoding from your source files using the H.264 codec, which has as close to universal device support as is currently possible. At the bare minimum, provide links to mobile-friendly files alongside any embedded video. Better yet, use a fully HTML5-compliant embed of an H.264 video.

The use of Flash for navigation or layout elements in a way that degrades appropriately for mobile devices or other platforms without Flash has, to put it mildly, a high degree of difficulty. Unless you've got some serious Flash development talent on staff or on retainer, it's best avoided.

The Nuclear Option

If we're honest with ourselves, an assessment of any library home pages will reveal links, sections, images, or features that are not strictly utilitarian. If you've got extraneous content on your site and are seeking to simplify for mobile use without redesigning, Cascading Style Sheets offer a simple way to nuke chunks of your pages. Similar to the method for designating a separate style sheet for print views of your pages, you can specify a mobile style sheet using the "media" attribute of the "link" tag when embedding your CSS like so:

```
<link rel="stylesheet" href="your_mobile_
stylesheet.css" media="handheld, only screen
and (max-device-width: 480px)" />
```

The above use of the "media" attribute indicates that this style sheet is to be used when a browser accessing the page is designed to use the "handheld" style sheet or has a screen of sufficiently low resolution. This technique will effectively capture most mobile devices, although newer high-resolution displays may require more targeted coding.

Once you've pointed your mobile users to their own style sheet, simply identify the extraneous bits of your pages that can effectively be removed. Give those divs the CSS "display:none" property, and they won't appear for anyone using a mobile device. This technique is crude, but if your home page is dominated by a Flash slideshow that your director is in love with, it

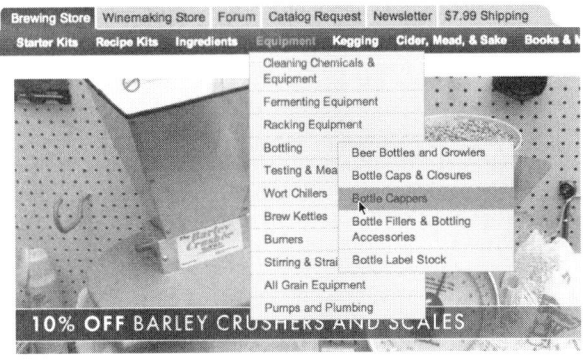

Figure 26
While mouseovers are a good feature on a standard website, they are problematic on mobile devices.

may be your easiest option for accommodating mobile users without much custom coding.

Develop a Mobile Website

If you have a little more time and resources to devote to your library's mobile strategy, consider developing a basic mobile website. It doesn't need to be anything fancy in order for it to be useful. By repurposing some basic content and linking to third-party mobile tools, you can likely cobble together a site that will be just what your users need when they want to access your library info on the go.

Designing and Developing Your Mobile Site

You can build mobile webpages just as you would any other webpage, using nothing more than a text editor. You will have the best results if you rely on web standards and build your pages using current versions of HTML, XHTML, and CSS. If you're familiar with jQuery or other JavaScript frameworks, there are a number of options for creating robust interactive sites with JavaScript, including the recently released jQuery Mobile.

jQuery Mobile: Demos and Documentation
http://jquerymobile.com/demos/1.0a1

Many of the leading mobile device manufacturers offer developer documentation on best practices for web applications. If you want a quick overview of design principles that apply to many mobile devices, I suggest HP/Palm's documentation "Developing Web Content for the HP webOS Platform." For an exhaustive reference on best practices for iOS devices, consult Apple's iOS Human Interface Guidelines.

In perusing Apple, HP, Android, BlackBerry, and Windows Phone 7 developer documentation, a few basic best practices become clear.

320 x 480 Layout

While the iPhone 4 and some newer Android handsets have a higher-resolution screen, sticking to 320 x 480 (figure 27) will ensure that your site is fully functional with the installed base of older smartphone models.

Viewport

Define a viewport in your site's meta tags that matches your mobile site's layout. The viewport sets the initial level of zoom for your site in the mobile browser. Smartphone web browsers are packed with features intended to make the nonmobile-optimized Web usable, and among them is a default viewport of nearly 1,000 pixels, designed to show the entirety of a website's width on initial load (figure 27). If you've designed a 320 x 480 pixel site and fail to set a matching viewport, your site will appear to have been shrunken to a third its actual size when initially loaded.

15-Point Text

The specific recommendations vary by manufacturer, but the average seems to hover around 15 points as the minimum size for readable text on a mobile webpage.

The screen is smaller than you think: If you're using an emulator on a standard computer monitor for development, you're seeing your designs at a much lower pixel density than on a mobile device.

Redirect Sparingly

Using packages like WURFL (the Wireless Universal Resource File), it's possible to detect specific platform/browser combinations on your web server and redirect them to your mobile site. Keep in mind that many of today's mobile devices can render nonmobile sites, and your users may prefer to stick with the site they're accustomed to. Consider redirecting users to an

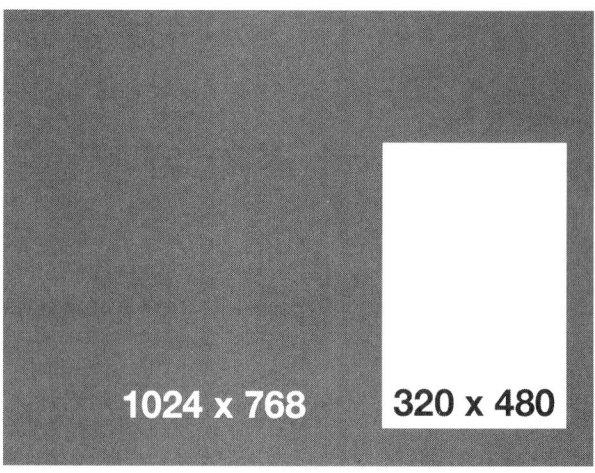

Figure 27
Viewport variation between the desktop and mobile browsers.

intermediary page where you give them the option of going to the mobile site or the main site (figure 28). Set a cookie to remember their preference.

Link to Your Main Site

Always provide a link to your nonmobile site. Not all of the functions of your main site may be available on your mobile site. Provide a path back to your main site in the footer of your mobile home page.

Local Content

Our experience with mobile web development at the University of Minnesota Libraries echoes what I've heard anecdotally from a number of colleagues at other institutions: the pages we use to serve up library hours information to mobile users are our most popular mobile pages by far.

I mention hours not just because I think it's important to make your hours easily accessible to your mobile users, but because it's instructive for considering the kind of information you serve mobile users and the contexts in which they're accessing said info. Users opt for accessing our site via mobile device when they are seeking small, concrete bits of task-specific information. In a rush to make it to the library? If you're like me, you're far more likely to risk tripping over a curb while finding hours or location information on a mobile device en route than you are to take the time to find the information using a laptop or desktop computer.

To quote from an earlier version of Apple's Human

Interface Guidelines:

> iPhone users are accepting of, and even anticipating, an experience different from the one they are accustomed to on a desktop computer, a laptop, or even a mobile phone. Although this affords you a certain latitude for experimentation, be aware that iPhone users are likely to be even less tolerant of sluggish performance and a complicated user interface than they are when it comes to software running in a computer.[1]

So, even though the University of Minnesota Libraries offer a fairly robust custom mobile search interface for our catalog and quite a few mobile databases, it's the hours and basic contact and location information that take the top slots for mobile users (figure 29). It's easy to fret about the complexity of creating mobile skins for our OPACs and federated search tools, but when it comes down to it, the most important information is relatively flat, text-based, and under our local control.

Vendor and Third-Party Mobile Interfaces

The number of vendors and aggregators providing mobile interfaces for their products has increased dramatically in the past couple of years. Whether you know it or not, your library likely already subscribes to one or more databases that already have mobile interfaces. Talk to your vendor reps and begin keeping track of which resources provide mobile web interfaces. You may be able to provide your mobile users a great deal of functionality simply by taking advantage of the work your vendors have already done for you.

Don't have the time, money or expertise to develop or purchase a mobile front end for your catalog? Consider linking to WorldCat Mobile (figure 30), where your users can answer the type of question they're likely to have on the go: does a resource exist that meets my need?

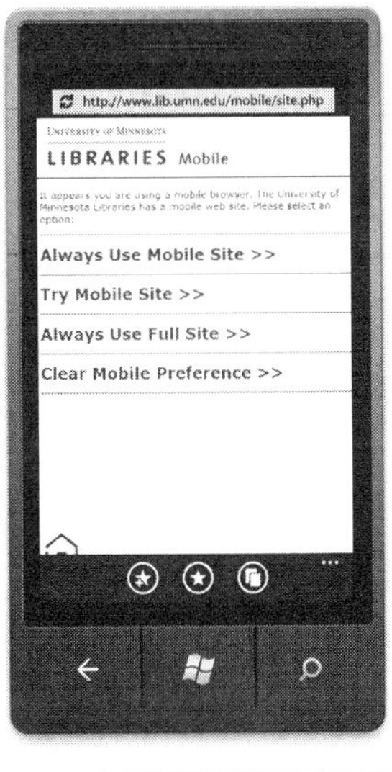

Figure 28
The University of Minnesota Libraries' mobile site allows users to set a preference for the mobile site or the full site.

Directory and Pages	Visits	Pageviews
/mobile	74,340	140,300
hours	16,492	26,591
site.php	15,700	21,567
staff	9,859	28,035
/	8,059	11,015
choice.php	3,557	4,364
search.php	3,029	4,989
primo.php	1,743	3,655
metasearch.php	1,731	2,480
index.php	1,600	2,732
indexes_az.php	1,531	2,159
Totals:	74,340	140,300

Figure 29
Stats from University of Minnesota's mobile site.

WorldCat Mobile Web Beta
www.worldcatmobile.org

If you're an Ebsco subscriber, you can direct your users to mobile versions of your Ebsco databases (figure 31), each of which will provide mobile-optimized access all the way to full-text PDFs, which should be viewable on most current smartphones.

PubMed has offered a spartan, but thorough, mobile interface for years, and it is quite usable on today's smartphones.

If your library uses the QuestionPoint or LibraryH31p for chat reference, each offers a version of its chat widget that is accessible by mobile devices.

Develop Mobile Web Applications

If you have access to sufficient developer expertise, you may consider adding levels of functionality to your mobile website by creating custom interfaces for your local systems, such as your OPAC, federated search, ILL, and circulation/account system. Increasingly these systems offer application programming interfaces (APIs) that allow for programmatic interaction with other applications. Such APIs in an OPAC or ILS might, for instance, allow you to take the input from a simple web form and pass it to the OPAC as a search query, then receive the results of that search back as a data stream that you can skin appropriately for your mobile web application.

You may be able to craft a similarly mobile-optimized interface for a link resolver, allowing users to search or browse for electronic journal titles. At the University of Minnesota, our developer John Barneson realized that while publishers and other vendors didn't consistently offer mobile versions of their sites, our federated search tool had a robust API, allowing him to create a mobile-optimized search interface not only for traditional federated searching, but for searches targeted to single databases. Using the API meant that we could, in one fell swoop, have a mobile search interface for any of our databases that were compatible with our federated search tool (figure 32).

When libraries begin to devote the time and resources necessary to create these more complex mobile tools, it's important to spend some time thinking about not just what is possible, but what's truly appropriate to present to mobile users. It is possible to direct users to Ebsco's mobile interface, or even to bring

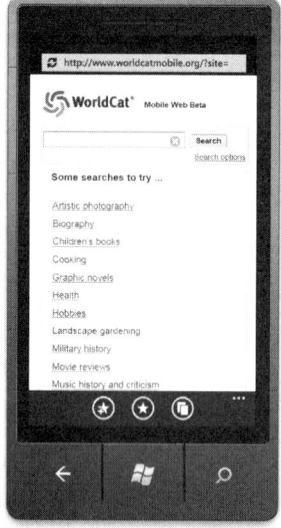

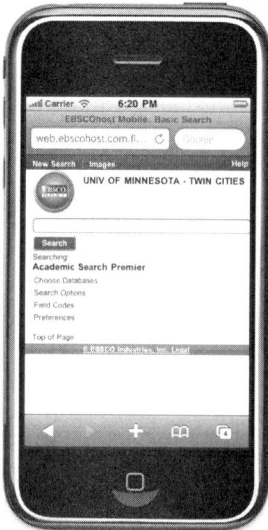

Figure 30
WorldCat's mobile interface.

Figure 31
Ebsco's mobile app.

them directly into the database via a mobile federated search interface. And once there, it is possible, on the vast majority of smartphones, to open up PDF files containing the full text of articles. However, the prospect of reading these PDFs on most mobile devices is still quite unappealing. The layout of most PDFs has been optimized for 8½-by-11-inch paper in portrait orientation, and, let's be honest, they're a pain to read on an average laptop screen, much less a mobile screen that may be as small as one tenth the size.

So, reading full-text PDFs is possible, but not necessarily appropriate. In considering common contexts of mobile device use, it's more appropriate to think of the user not in the mindset of concerted researching, but instead in brief moments of verification. Rather than working to support complete digestion of an information object, we should optimize our web applications around discovery and, most importantly, marking or holding for later retrieval. The appropriate calls to action in a mobile library web application should be "bookmark this," "request this item," "e-mail me these records," and so on.

There is one gee-whiz mobile web application feature that I'd like to see more libraries implement. Most smartphone browsers support the draft HTML5 Geolocation API, which allows the user to grant webpages access to the phone's location information. Likewise, many licensed electronic resource agreements allow for free public access on site in the library. How cool would it be to create your library web application such that users could be authenticated into your electronic resources on their mobile devices by proving that they were in your library via geolocation? Very cool is the answer.

W3C's Geolocation API Specification
http://dev.w3.org/geo/api/spec-source.html

Develop Native Mobile Apps (or Don't)

As mentioned earlier in this report, the iOS App Store for iPhone, iPod Touch, and iPad has been a juggernaut. In the two and a half years since the App Store opened its virtual doors within Apple's iTunes media management and syncing application, more than 250,000 applications have become available, and have been downloaded over 6.5 billion times. Developers of paid iOS applications split the sale price 70/30 with Apple, and by mid-2010 the 70 percent that Apple pays developers had topped $1 billion.[2]

The Android Marketplace has lagged behind the iOS App Store in the number of applications available for download, the number of applications downloaded by users, and the profits realized by developers. However, as Android adoption nears or surpasses that of iOS, the Android development scene has begun to blossom.

Apps have truly become the watchword of this new mobile renaissance. But you probably shouldn't build one.

Let me explain. There are three reasons why you might want to build a native mobile app:

1. Profit: It's clear that it's possible to make money building software for mobile devices, especially for iOS.
2. Native device functionality: If you have an idea for an application that requires access to the camera, the compass, the tilt sensor, you won't be able to do that with a web app.
3. Exposure: You may assume that simply having an app will burnish your reputation among the cutting edge and provide opportunities for press releases.

So, to examine each of these reasons. First, if you manage to figure out a way for your library to turn a profit with a native app other than by jamming it full of advertising, call me. With some notable exceptions,

Figure 32
Searching a database.

we're not in the business of making a profit, nor for charging for our services. So, unless you have a very unusual library or some very unusual collections that would lend themselves both to mobile applications and to monetization, it's unlikely that profit is an appropriate motivation for building an app.

Implementing native device functionality can be very, very cool. Apps like RedLaser (owned by eBay) use the iPhone or iPod Touch camera to scan UPC barcodes so you can compare prices on items across various online resellers. They've also partnered with WorldCat so that when you scan a book you can see if it's available in a local library. RedLaser even provides an SDK for its barcode scanning function, meaning a library could craft its own app using this technology. Functionality like this is simply not available through anything other than a native app.

There's no doubt that having mobile apps provides an opportunity to talk up the cutting-edge developments at your library. However, it's no longer the case that simply having an app will drive users to your library. Between the various smartphone platforms, there are nearly half a million applications available, and a library application with local or regional reach is unlikely to make it into the "top download" charts that drive massive adoption.

There is one overarching reason why you might not want to build a native mobile app: If you want to do it well, it's expensive.

Craig Hockenberry, the developer of the popular iOS Twitter client Twitterrific recently responded online to a programmer who speculated that the app likely took 160 hours of development time and 40 hours of graphic design time (plus another month for testing). Hockenberry implied that estimate must have been generated under the influence of illicit substances. A recent update to the program had in fact taken more than 1,100 developer hours and 225 designer hours. He estimated the total cost at nearly a quarter of a million dollars in time and expenses.[3]

Now, Twitterrific is a highly polished application, but at its core, it's simply a mobile-native interface for Twitter's API. It's not that much different conceptually from what a library's native app might look like,

except that a library's app would likely be more complicated and involve creating an interface for a number of different systems' APIs. And Twitterrific is only for iOS. Developing for Android, BlackBerry, webOS, or Windows Phone 7 would require a complete re-creation of the code, as these platforms share almost no common elements.

It is certainly possible to create native applications on the cheap. There are some tools like PhoneGap that provide options for building a single application that can be deployed to multiple mobile platforms. If you already have a mobile website, it's possible to package it in an application that acts as a simplified web browser dedicated to your site. However, in doing so you're asking your users to take the additional steps of downloading and installing an application and are offering no additional functionality to justify their effort. Worse, unless you devote developer time to creating data that can be stored locally on the users' mobile device, you've created an application that won't work offline, one of the key differentiating factors of apps versus the Web.

PhoneGap
www.phonegap.com

Notes

1. Apple Developer Connection, "It's a Browser-Based World," 2008, www.interactivefish.com/apple/section1.php (accessed Jan. 4, 2011).
2. Apple, "Statement by Apple on App Store Review Guidelines" (press release), Sept. 9, 2010, Apple Press Release Library, www.apple.com/pr/library/2010/09/09statement.html (accessed Jan. 5, 2011).
3. Craig Hockenberry, answer posted Oct. 13, 2010, to "How Much Does It Cost to Develop an iPhone Application?" posted Oct. 16, 20008, by user27815, http://stackoverflow.com/questions/209170/how-much-does-it-cost-to-develop-an-iphone-application/3926493#3926493 (accessed Jan. 4, 2011).

Chapter 4

Issues for Information Access on the Mobile Web

Abstract

Although offering mobile services is a tremendous opportunity to expand a library's ability to provide service to its community, the mobile Web is still a new technology and a seamless transition into mobile services is practically impossible. This chapter of Libraries and Mobile Services *examines the most common issues and problems that librarians will face.*

The mobile Web is a heady place. Users are presented every day with new, more efficient, and better ways to access information anywhere at any time. Library professionals should be excited about the opportunities presented by the rapidly increasing pace of mobile technology evolution. There are also some issues that deserve our ongoing scrutiny as stewards of information and access.

Security

To a certain extent we've all become accustomed to the security issues that face us when using a Net-connected desktop or laptop computer. We know not to click links in suspicious e-mails. We've learned that no matter what an e-mail says, Nigeria has been a republic since 1999 and has not in modern history had a royal family any member of which needs your assistance to liberate their fortune. Unfortunately, we need to anticipate these same security concerns on the mobile Web, though perhaps with more disturbing consequences.

Recently, audits of the Android Marketplace have found dozens of applications that seem to do little but harvest information from users' phones. Masquerading as wallpapers or themes, these apps ask users for broad permissions to access the phone's file system. Unfortunately, even well-behaved Android apps also require this kind of permission, so it can be hard to distinguish the bad actors in the marketplace.

The growth in mobile payment applications like PayPal and Square, combined with social networks like Facebook and location-based services like foursquare, mean that the information available to hackers on mobile devices is potentially very damaging.

First Sale

The bread and butter of libraries for many years has been the first sale doctrine. This is what allows us to purchase books, CDs, and DVDs and then lend them over and over again. The first sale doctrine dictates that creators and publishers have a right to make a profit on the first sale only of their works, and that they have no claim over the proceeds of subsequent sales. We've seen for years now in libraries as we transition from physical to digital materials that first sale no longer applies.

In the mobile realm, first sale barely applies to the devices themselves, which often come with a two-year contract obligation. All the applications, e-books, music, and video users are purchasing and using on their mobile devices are licensed, not owned. For their purchase price, users are granted the right to use the digital file on a limited number of devices and under a specific set of circumstances, as defined by a fine-print End User License Agreement. Libraries may not, under the terms of these licenses, redistribute Kindle e-books or a copy of the Angry Birds game, or a video from the iTunes Store.

Cost

While the number and demographics of people who have Internet access via a mobile device are inspiring for anyone who has been keeping track of digital divide issues, the economics of mobile Internet access are frustrating. One reason mobile carriers have so eagerly championed smartphones is that they are able to charge an additional monthly fee on top of the voice plans they offer users. Typically these fees add $20–30 per month to the phone bill per user.

Bandwidth on mobile devices tends to be a fraction of that available to users on standard home cable or DSL service. This has improved in recent years and will likely continue to improve as carriers upgrade to 4G LTE or WiMAX service, but those upgrades will require handset upgrades as well, and may come at an additional monthly cost.

In addition, while most home Internet plans are limited only by bandwidth, meaning the amount of data you can pull through your connection at any given time, most mobile data plans are limited to a set amount of data traffic per month. This means that once you've streamed a couple of gigabytes worth of movies (not a difficult task, especially if your mobile device is your primary entertainment device), you're either (a) cut off for the rest of the month, (b) downgraded to a very slow connection for the rest of the month, or (c) charged a per-kilobyte overage fee for the additional data you use in that month.

Coverage

While most of us who live in metropolitan areas can choose from a number of mobile providers, and as a result from a number of handsets, mobile operating systems, and data plans, there are still many areas in the country where there is little mobile coverage or where choice between providers may be limited or nonexistent. Figure 33 shows areas of the United States identified by the FCC according to the number of mobile providers available to consumers. Libraries have been vocal advocates for rural broadband penetration and ought to be similarly strong advocates for the availability of competitive mobile data.

Net Neutrality

Net neutrality is shorthand for the principle that the Internet is a level playing field for any and all data traffic. That is to say that my Internet service provider should treat bits flowing into my browser from one source exactly the same way as those from another. Many would argue that this idea is part of the foundation of the Internet. It was taken for granted for many

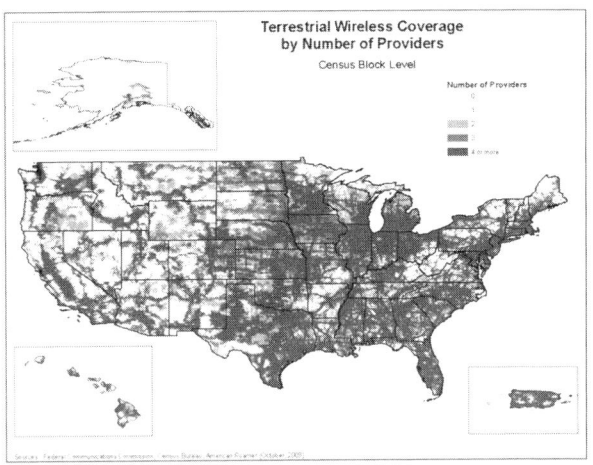

Figure 33
FCC data showing number of mobile providers by area.

years until recently, when online delivery of content, especially multimedia content, became big business. It's in the age of streaming web video that entrenched interests have found themselves at odds.

If you are like me, you get your home Internet from a cable company. I happen to enjoy several streaming video services, like Netflix and Hulu, which provide television shows via the Internet, which I get through a cable company. What I don't get from the cable company is cable TV. If you view this transaction from the perspective of the cable company boardroom, what you see is Netflix and Hulu using the cable company's own pipes to steal its business. Net neutrality dictates that the cable company has to deal with the situation. They must deliver Netflix's bits to my computer just as fast as anything else, and without additional fees, because they're just bits.

While net neutrality has largely held when it comes to standard wired Internet service, it is under direct attack in the mobile world. Google and Verizon, sick of waiting for an FCC ruling on wireless net neutrality, made their own pact. This proposal lays out in no uncertain terms what exactly net neutrality means for wired Internet service, and then goes on to state that wireless networks should be exempt from almost all of it.

The New Digital Divide

The unfortunate truth is that while more people than ever before will have Internet access because of its availability through mobile devices, there's a very real possibility that said Internet access will be the most expensive, slow, and restricted access since WebTV. For the same cost as a single mobile data plan, a family could have DSL access that would serve all the family members with greater bandwidth and greater assurance of ongoing net neutrality.

Conclusion: The Future of Mobile Computing and Libraries

As I write this, my wife is pregnant with our first child. Over the past few months, it's the only thing that I've spent more time thinking about than mobile technology. I suppose some overlap in my ruminations was inevitable.

A few weeks ago I began to think about what the mobile future held for my son or daughter. I recalled when I first began to use communication technology independently, using my family's wired home telephone to contact a friend on his family's wired home telephone. This scenario had remained unchanged since my parents were children. I realized that I have absolutely no idea how my child will communicate with his or her peers.

I haven't had a proper home telephone for ten years. The phone that I use most frequently I use far more often as a web browser, e-mail terminal, instant messenger, game console, social networking hub, or media player than I do as a phone. For all the time I've spent obsessing about communication technology, I can't for the life of me envision how my child will invite another child over to play. This is to say nothing of how he or she will access news, entertainment, or other media.

There are aspects of the evolution of mobile computing that are very easy to predict:

- Screen resolution and fidelity will continue to increase, providing a more pleasant reading and media viewing experience.
- Battery technology will improve, giving us longer life in our devices, even as we demand more features from them.
- Processor speed and efficiency will grow, allowing mobile devices to approach, and eventually surpass, a level of computing power that we currently associate with desktop computers and servers.
- Mobile bandwidth will increase in both availability and speed.

All of these technologies are effectively commoditized, and their continued evolution is all but assured as described by concepts like Moore's Law. Guessing at what exactly these technological advancements will mean when taken together and paired with the software and services they enable is far more challenging.

If I have one prediction about the future of mobile computing it's this: the future of mobile is the future of computing.

In the coming years the differences between the computers sitting on our desks and the computers we carry in our pockets will disappear rapidly. People have long anticipated the day when we have mobile devices that are the equivalent of laptops in our pockets. I think this is backward. I think laptops and desktops have much more to learn from mobile than mobile does from traditional computers.

The constraints inherent in mobile devices have provided an opportunity to do away with decades of presuppositions about how we interact with computers and with information, unleashing creative and compelling new user experiences that have quickly surpassed the desktop.

Unfortunately, those same presuppositions underlie many of the decisions libraries have made about how we provide services to users. The cloud-based streaming and licensed content models that are revolutionizing mobile content delivery, such as Pandora, Kindle, the iOS App Store, and Netflix, are fundamentally incompatible with the model of library lending. This is unlikely to change.

If mobile is the future of computing, libraries must think of mobile first in all digital services, whether homegrown or licensed. We must use platform-agnostic web standards, and demand the same from our vendors. We must advocate as tirelessly for mobile broadband access and net neutrality as we have for wired broadband.

I'm not sure how my child will communicate with his or her friends. I have no idea whether my child will learn to read off of a printed page or off of a screen. I don't know whether visiting the library will mean hopping on our bikes or using a mobile device. But with your help, visit we shall.

Notes

Library Technology Reports Respond to Your Library's Digital Dilemmas

Eight times per year, *Library Technology Reports* (*LTR*) provides library professionals with insightful elucidation, covering the technology and technological issues the library world grapples with on a daily basis in the information age.

Library Technology Reports 2011, Vol. 47	
January 47:1	**"Web Scale Discovery Services"** by Jason Vaughan
February/ March 47:2	**"Libraries and Mobile Services"** by Cody W. Hanson
April 47:3	**"Using WordPress as a Library Content Management System"** by Kyle M. L. Jones and Polly Alida-Farrington
May/June 47:4	**"Perceptions of Library Automation: An International Study"** by Marshall Breeding and Andromeda Yelton
July 47:5	**"Using Web Analytics in the Library"** by Kate Marek
August/ September 47:6	**"Re-thinking the Single Search Box"** by Andrew Nagy
October 47:7	**"The Transforming Public Library Technology Infrastructure"** by ALA Office for Research and Statistics
November/ December 47:8	**"RFID In Libraries"** by Lori Bowen-Ayre

ALA TechSource

alatechsource.org

ALA TechSource, a unit of the publishing department of the American Library Association